Sensors

Kourosh Kalantar-zadeh

Sensors

An Introductory Course

Springer

Kourosh Kalantar-zadeh
School of Electrical and Computer Engineering
RMIT University
Melbourne, Victoria, Australia

ISBN 978-1-4899-9984-9   ISBN 978-1-4614-5052-8 (eBook)
DOI 10.1007/978-1-4614-5052-8
Springer New York Heidelberg Dordrecht London

Printed on acid-free paper

Springer is part of Springer Science+Business Media (www.springer.com)

To Stephanie

Preface

The basic contents for many of the university courses were established in the 1960s. While these courses are still dynamic and lecturers continually add new materials to the curricula, for many of them the fundamentals have remained the same. However, when it came to teaching courses such as sensors, the situation was completely different. The advancement of the sensors field in the early 2000s, due to the emergence of new materials and devices, has been staggering. However, there was an absence of a clear reference text that could both provide the fundamentals of this multidisciplinary field and also stimulate student curiosity and imagination.

In 2006–2007, I coauthored a book entitled "Nanotechnology Enabled Sensors" (published by Springer). The book has many chapters that are exclusively about sensors but the focus of the text was on devices strongly influenced by nanotechnology. As a result, it was not a suitable reference book for a course only on sensors. In addition, the book still requires a comprehensive editorial revision. Consequently in 2009, I drew my attention to writing this book entitled "Sensors: An Introductory Course". The book still contains many sections borrowed from "Nanotechnology-Enabled Sensors"; however, the technical parts, which were previously focused on nanotechnology, have been removed and many pages on the basics of sensors and their operations have been added. The aim of this book is to provide an easy-to-understand and engaging text on sensors for university students, or those entering the field, with a large number of examples and well-designed illustrations.

The preparation of the materials for this book was a long, and at the same time, rewarding process. As the textbook subject was multidisciplinary, I needed to ensure a thorough understanding of students' background knowledge in the relevant disciplines. Consequently, I spent many hours researching materials presented to students from different disciplines and investigated learning resources about sensors from the point of view of biotechnologists, analytical chemists, physicists, medical doctors, and engineers. Visiting many laboratories and familiarizing myself with different analytical tools that are conventionally used in measurements and sensing were also important parts of this task. I carefully studied over 100 textbooks and 2,000 scientific papers to ensure the full gamut of devices, which are implemented in sensing systems, was included. In addition, I studied the literature on pedagogy and also sat in many classes of relevant disciplines and engaged in discussions with the students of these disciplines in order to learn their point of view about educational engagement and learning. To ensure that the learning materials were relevant to the real-world experiences students would have in the field and were connected to what they would encounter and implement after their graduation, I visited numerous research and development firms, industrial factories, offices, and hospitals and identified the types of sensors that are commonly implemented in such places as well as the ways they are utilized and maintained.

The other important part of the editing process was started after the preparation of the first draft. At this stage, I actively sought feedback from students of chemistry, physics, biotechnology, and engineering to appreciate their point of view and ensure that the materials presented were easy to understand, regardless of the students' background knowledge, and would meet their learning needs. After this stage, the text was passed to approximately ten different academics, who provided the final comments that I meticulously incorporated into the final version of the book.

The text was designed in a way to encourage students to familiarize themselves with the real-world applications of sensors. All efforts were in place to assure that there was a clear connection between the student learning materials and what is happening within the wider research and industrial community in the field of sensors.

Melbourne, VIC, Australia   Kourosh Kalantar-zadeh

Contents

Chapter 1
Introduction

Abstract In this chapter a general overview regarding sensors will be presented. Some of the fundamental terminologies, which are frequently encountered in the sensor field, will be described. The importance and applications of sensors will be highlighted.

1.1 Sensors and Transducers

The word *sensor* has a Latin root "sentire," which means "to perceive," originated in 1350–1400, the Middle English era. A sensor is a device which responds to stimuli—or an input quality—by generating processable outputs. These outputs are functionally related to the input stimuli which are generally referred to as *measurands*.

Transducer is the other term that is sometimes interchangeably used instead of the term sensor, although there are subtle differences. A transducer is a device that converts one type of energy to another. The origin is "transduce," which means "to transfer" that was first coined in 1525–1535. A transducer is a term that can be used for the definition of many devices such as sensors, actuators, or transistors.

Schematic diagram of a sensor is depicted in Fig. 1.1. A sensor is commonly made of two major components: a *sensitive element* and a *transducer*. The sensitive element has the capability to interact with a target measurand and cause a change in the operation of the transducer. Affected by this change, the transducer produces a signal, which is translated into readable information by a data acquisition system. It is important to note that for devices such as transistors and actuators, the *conversion efficiency* is generally an important factor. However, in addition to the conversion efficiency, a sensor should have many other qualities such as *selectivity* and *sensitivity*, which will be explained later.

Other common definition is to use the term sensor for the *sensing element* itself and transducer for the *sensing element plus any associated peripherals* (the overall system). For example, a temperature sensor is called a sensor, while together with the data acquisition circuit (to convert the signal into a measurable electrical voltage) is called a transducer.

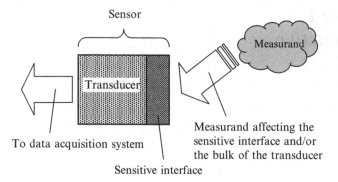

Fig. 1.1 Schematic depiction of a sensing system

Most of the man-made systems acquire data from their surrounding environments, process them, and consequently translate the data into useful functions. Sensors' roles in such systems are the acquisition of data. Using sensors, direct contact with the surrounding world is more possible. Sensors enable systems to interact with their environments and receive desired information. This information is then sent to a processing system, where it is processed into meaningful information. The processed information can be either directly deciphered as the sought after output or further fed into another system for additional processing or producing signals for actuators. The more complex the system, the larger number of sensors is required for its operation.

1.2 Sensors Qualities

Sensors should be *sensitive* to their target measurands, and *insensitive* to any other input quantities, which might impact on their performance. As a universal rule, sensors should provide *reliable, accurate, stable,* and generally *low cost* sensing.

A good sensor *should not affect* the measurand. For instance, if a sensor is used for measuring temperature, the size of the sensor should be much smaller than size of the monitored object (e.g., micro-/nano-sized objects) or the measurement should be conducted remotely (infrared temperature measuring systems). A sensor *life time* is important—some should last for a long time—such as smoke detectors for houses—some are disposable—such as pregnancy test sensors (which should be kept fresh and intact in sealed packaging until their applications). *Response behavior* of a senor should be well-understood by users: parameters such as *linearity* and *repeatability*, which will be discussed in detail in Chap. 2. These parameters are presented in the *data sheet* of a sensor and a good sensor must have a comprehensive and clear data sheet. A senor has to be *well calibrated* before the application. This means an extra cost for the manufacturer. *Interferences* of signals other than the measurand should be minimal for a good sensor. Sensors *resolution*, which will also be discussed in Chap. 2, should be well presented to the consumers via the data sheet.

A sensor's *fabrication cost* should be as small as possible. Economy always governs the engineering and design, and a lower cost is always a winner. A sensor should be *environmentally friendly*. Many types of sensors are disposable and all sensors have limited life span. After the end of their lives, they should be able to be return to nature safely. For instance, in the past many sensors were using mercury in their structure, which are now mostly banned from the manufacturing. *Selectivity* of the sensor is also an important issue. A good sensor is selective to the target measurand. A discussion on selectivity will also be presented in Chap. 2. In addition, a sensor's *power consumption* should be manageable and always considered in the design and fabrication. One of the main hurdles of *sensor networks* is the sensors', and their circuits, demand for power. In a network, sensors can be scattered all over a field and they should be individually supplied with sufficient power. This always poses a challenge. Stand-alone sources, such as batteries, have limited lifetime. Renewable sources, such as solar cells, have their own complexity of operation and add to the costs. Providing the energy form a central source requires expensive and complicated wiring matrix. Another issue is the communications between the central data acquisition system and sensors in the nodes of such sensor networks. The discussion on sensor networks is outside the contents of this book though and the author advice readers to refer to excellent available textbooks in the field.

1.3 Types of Transducers

Since the *conversion of energy from one form to another* is an essential characteristic that governs the sensing process, it is important to be familiar with different forms of energies. The list is presented in Table 1.1.

A general schematic of a sensing system is shown in Fig. 1.2. As can be seen, the input energy is entered via an input interface, goes through a transduction process, and is released as a signal processable for a user via an output interface.

Table 1.1 Various forms of energies and their occurrence

Type of energy	Occurrence
Gravitational	Gravitational attraction
Mechanical	Motion, displacement, mechanical forces, etc.
Thermal	Thermal energy of an object increases with temperature. In thermodynamics, thermal energy is the internal energy present in a system in a state of thermodynamic equilibrium because of its temperature
Electromagnetic	Electric charge, electric current, magnetism, electromagnetic wave energy (including the high frequency waves such as infrared, visible, UV, etc.)
Chemical	Energy released or absorbed during chemical reactions
Nuclear	Binding energy between nuclei—binds the subatomic particles of a matter

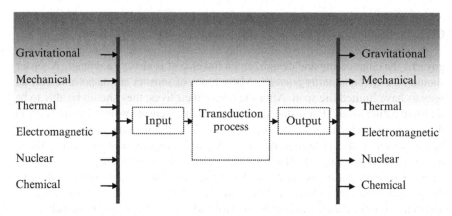

Fig. 1.2 Schematic depiction of a sensing system interacting with various possible input and output energies

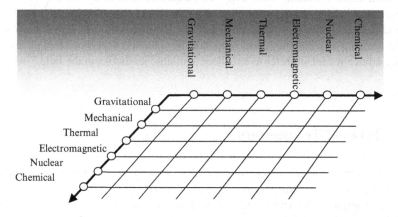

Fig. 1.3 A two-dimensional representation of possible direct transductions from one form to another

A two-dimensional representation of the direct transduction, from one form of energy to another, is shown in Fig. 1.3. The primary energy input to the system is represented by one axis and the energy output by the other axis. With the 6 forms of energy, we have $6 \times 6 = 36$ possibilities.

Physical and *chemical effects* are involved in signal transductions. *Physical effects* involve those that couple a material's thermal, mechanical, electromagnetic (including optical), gravitational, and nuclear properties. These effects together with the *chemical effects* are utilized within sensors. Table 1.2 shows examples of effects that are obtained when these properties are coupled with each other or with themselves.

Table 1.2 Examples of some physical and chemical effects

	Nuclear	Gravitational	Thermal	Mechanical	Electromagnetic (including optical)	Chemical
Thermal	Fusion, fission	–	Heat transfer	Thermal expansion	Thermoresistance	Endothermic reaction
Mechanical	–	–	Friction	Acoustic effects in musical instruments	Magnetostriction	–
Electromagnetic (including optical)	Synchrotron	–	Peltier effect	Piezoelectricity	Hall and Faraday effects	Electrodeposition
Chemical	Nuclear reactions	–	Exothermic reaction	Heat engines	Batteries and fuel cells	Chemical interactions
Gravitational	Sun power	–	–	–	Acceleration by gravity	–
Nuclear	Nuclear reactions	–	Nuclear power plants	Shock waves in nuclear explosion	–	–

A transduction system can be more complicated than the ones shown in Fig. 1.2. The procedure might consist of the transduction of energy via *several consecutive steps*. For instance, the input signal can be in the form of a mechanical stimulant, which generates heat (e.g., via friction) and this heat is then transformed into a voltage (electromagnetic energy), which is a measurable signal for a user. In this case, we have an extra intermediate transduction process. This adds extra dimensions to the transduction plane representation of Fig. 1.3. In this case an extra intermediate step, we are dealing with a space with $6 \times 6 \times 6$ possibilities. Surely for more complicated systems the number of possibilities increases, if the number of in-between transduction steps increases. Such systems can be more complex and costly but in return can increase the reliability and performance.

1.4   Sensors Applications

In recent years the development of operational systems has become progressively important and sensors have become an ever-present part of such systems. Sensors are playing a greater than ever role in our day-to-day interactions and are becoming an integral part of the modern technological growth and development. Each application places various requirements on a sensor and its integrated sensing system. However, regardless of the type of application all sensors have the same object: to achieve accurate and stable monitoring of target measurands.

Sensor technology has flourished as the need for physical, chemical, and biological recognition has grown. Nowadays sensors are finding a more prominent role, as strong needs on devices aimed at making lives better, easier, and safer are observed. They are employed in applications ranging from environmental monitoring, medical diagnostics and health care, automotive and industrial manufacturing, home appliance, defense and security, and even toys. In recent times, however, the importance of sensors has grown significantly due to increasing automation and more use of microelectronics, both of which require more sensors. Parallel to this development, the capabilities of sensors are increasing and sensors prices are shrinking. Application of sensors can be categorized as follows:

- Health and biomedical.
- Defense and military industries.
- Industrial applications: aerospace, agriculture, nuclear, automation, automotive, transportations, building technology, machine control, power generation, textile, chemical, and food industries.
- Homeland security and safety.
- Environmental surveillance and climate parameters measurements.
- Consumer products: electronic systems and household appliances.

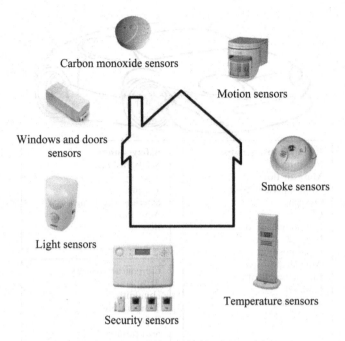

Fig. 1.4 Examples of sensors incorporated in a typical house

Sensors are found commonly around the household. For example, they are located in gas cook tops, where they determine whether or not the pilot is on, and if not, halt the gas flow preventing the room from being filled with gas. They are in electrical devices from surge protectors to automatic light switches, refrigerators and climate control appliances, toasters, and of course in smoke detectors (Fig. 1.4).

We encounter sensors in everyday life: entering a department store with automatically opening doors, or when light is automatically turned on and off upon entering or leaving an office.

Sensors are also an integral part of health care and diagnostics. They can determine whether or not biological systems are functioning correctly and most importantly, direct us to act without delay when something is wrong. For instance glucose meters are playing a crucial role in determining the amount of sugar levels in people diagnosed with diabetes. They are incorporated in off-the-shelf blood pressure and oxygen content monitoring as well as in pregnancy tests systems.

The functions of senors are generally not so noticeable, yet millions of them are contained within central processing units of computers and microcontrollers. In addition to these internally integrated electronic devices, electronic systems have a large number of external sensors for interacting with their users. Touch screen pads, mouse, and voice recognition systems all use sensors in their operations.

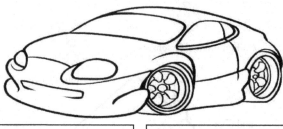

Essential drive sensors	Safety sensors
Pressure	Safety distance
Mass air flow	Tilt
Atmospheric pressure	Torque
Oxygen	Steering wheel angle
CO_2	Acceleration
Rotational speed	Belt
Petrol level	
Pedal position	**Convenience sensors**
Angular position	Air quality
Engine temperature	Humidity
Oil level	Temperature
Crankshaft position	Rain
	Seat positon

Fig. 1.5 Example of sensors incorporated into a typical vehicle

Nowadays almost all engineering machines incorporate sensors. Sensors are widely used in automotive industry. An average vehicle can have hundreds of individual sensors for different functions (Fig. 1.5). This includes: sensors for doors, wipers, several temperature and exhaust sensors around engine and exhaust pipes, and gas flow sensors. Aircraft are riddled with them as they monitor position, wind speed, air pressure, altitude, etc. Another important use is for industrial and process control, where the sensors continually monitor to ensure that efficiency is maximized, production costs are minimized, and that waste is reduced.

The choice of a senor for a specific task is a process that should be well-thought during the design and implementation process. The question of "which sensor should be used in what application" might have several answers, depending on the circumstances. For instance, for measuring "seat position in a car" a variety of different sensors such as Hall effect, capacitor, inductor, and magnetoresistor-based sensors can be used (the operation of these types of sensors will be explained in later chapters of this book). There is no limitation and no real rule in the choice of the most suitable sensor for this task. it can depend on many different parameters such as design of the sensors, design of the seat, the size of the vehicle, the availability of the power supply, the total cost of the vehicle, and many more other parameters. A good sensor engineer is the person who always finds the best solution considering the circumstances and opportunities.

1.5 About this Book

In this book, firstly the basic terminologies and fundamentals of sensor characteristics will be described. Physical transduction effects will be presented next and sensor templates will be explained and their peculiarities be highlighted. Organic sensors, with an emphasis on biosensors, will be presented in the final chapter and readers will become familiar with the operation of such sensors and their applications.

The sensor field is multidisciplinary by nature. Therefore, the book where it is needed, presents the basic background knowledge regarding the physics, chemistry, and biology of sensors.

The book will cover the most applied sensors in the market and will highlight their conventional and advanced applications. Technical information is followed by in-depth discussions and relevant practical examples in each chapter.

This book can be utilized as a text for students who are entering the field of sensors for the first time. It is written in a manner that early-year university students in the fields of chemistry, physics, electronics, biology, biotechnology, mechanical engineering, and bioengineering can benefit. It can also serve as a reference for engineers already working in this field.

Chapter 2
Sensors Characteristics

Abstract In this chapter, static and dynamic characteristics of sensing systems will be presented. Their importance will be highlighted and their influence on the operation of sensing systems will be described.

2.1 Introduction

After receiving signals from a sensor, these signals need to be processed. The acceptable and accurate process of these signals requires: (a) full knowledge regarding the operation of the sensors and nature of signals, (b) *posteriori knowledge* regarding the received signals, and (c) information about the *dynamic* and *static characteristics* of the sensing systems.

(a) In order to be able to use signals' information correctly, the operation of a sensor, and the nature of signals they produce, should be well understood. By having this knowledge, we are able to choose the right tools for the acquisition of data from the sensor. For instance, if the sensor output is voltage, we utilize analogue-to-digital and sample and hold circuits, as well as a circuit that transfers the digits into the computer. If a sensor produces a time varying signal where the information is embedded in its frequency signatures, then a frequency counter and possibly a frequency analyzer are needed. If output of the sensor is a change in color then a visible spectrometer is necessary.

(b) A posteriori knowledge (a posteriori knowledge or justification is dependent on experience or empirical evidence) about the received signals is important in order to assure that the data will be interpreted correctly and that the right device is used in the measurement process. We need to have a good understanding for what is expected from the sensor and system. For instance, even during a simple DC voltage reading, if the DC input has been mixed with AC signals (may happen often due to the influence of unwanted electromagnetic waves), the measured value can be significantly different from the real measurand.

K. Kalantar-zadeh, *Sensors: An Introductory Course*,
DOI 10.1007/978-1-4614-5052-8_2, © Springer Science+Business Media New York 2013

In this system, the presence of unwanted AC signals can produce unrealistic and meaningless measurements. If knowledge regarding the presence of AC voltages were available (a posteriori knowledge), a filtering process could be efficiently used even before feeding the stimuli into the sensing system (e.g., electromagnetic shielding or filter to remove 50 Hz AC signals). Having knowledge about the characteristics of sensing systems also allows us to extract meaningful conclusions with minimal error. For example, we can avoid wrong readings at short time brackets; if we know that a gas sensor needs 5 min to respond to a target gas rather than 5 s.

(c) The characteristics of a sensor can be classified into two *static* and *dynamic* groups. Understanding the dynamic and static characteristics behaviors are imperative in correctly mapping the output versus input of a system (measurand). In the following sections, the static and dynamic characteristics will be defined and their importance in sensing systems will be illustrated.

2.2 Static Characteristics

Static characteristics are those that can be measured after all transient effects have been stabilized to their final or steady state values. Static characteristics relate to issues such as how a sensor's output change in response to an input change, how selective the sensor is, how external or internal interferences can affect its response, and how stable the operation of a sensing system can be.

Several of the most important static characteristics are as follows:

2.2.1 Accuracy

Accuracy of a sensing system represents the correctness of its output in comparison to the actual value of a measurand. To assess the accuracy, either the system is benchmarked against a standard measurand or the output is compared with a measurement system with a superior accuracy.

For instance considering a temperature sensing system, when the real temperature is 20.0 °C, the system is more accurate, if it shows 20.1 °C rather than 21.0 °C.

2.2.2 Precision

Precision represents capacity of a sensing system to give the same reading when repetitively measuring the same measurand under the same conditions. The precision is a statistical parameter and can be assessed by the standard deviation (or variance) of a set of readings of the system for similar inputs.

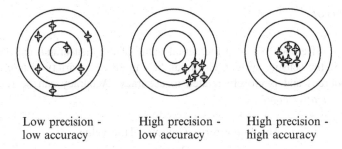

| Low precision - low accuracy | High precision - low accuracy | High precision - high accuracy |

Fig. 2.1 The difference between accuracy and precision

For instance, a temperature sensing system is precise, if when the ambient temperature is 21.0 °C and it shows 22.0, 22.1, or 21.9 °C in three different consecutive measurements. It is not considered precise, if it shows 21.5, 21.0, and 20.5 °C although the measured values are closer to the actual temperature. The game of darts can be used as another good example of the difference between the accuracy and precision definitions (as can be seen in Fig. 2.1).

2.2.3 Repeatability

When all operating and environmental conditions remain constant, *repeatability* is the sensing system's ability to produce the same response for successive measurements. Repeatability is closely related to precision. Both long-term and short-term repeatability estimates can be important for a sensing system.

For a temperature sensing system, when ambient temperature remains constant at 21.0 °C, if the system shows 21.0, 21.1, and 21.0 °C in 1 min intervals, and shows 22.0, 22.1, and 22.2 °C after 1 h, in similar 1 min intervals, the system has a good short-term and poor long-term repeatability.

2.2.4 Reproducibility

Reproducibility is the sensing system's ability to produce the same responses after measurement conditions have been altered.

For example, if a temperature sensing system shows similar responses; over a long time period, or when readings are performed by different operators, or at different laboratories, the system is reproducible.

2.2.5 Stability

Stability is a sensing system's ability to produce the same output value when measuring the same measurand over a period of time.

2.2.6 Error

Error is the difference between the actual value of the measurand and the value produced by the sensing system. Error can be caused by a variety of internal and external sources and is closely related to accuracy. Accuracy can be related to *absolute* or *relative error* as:

$$\text{Absolute error} = \text{Output} - \text{True value},$$
$$\text{Relative error} = \frac{\text{Output} - \text{True value}}{\text{True value}}. \tag{2.1}$$

For instance, in a temperature sensing system, if temperature is 21 °C and the system shows 21.1 °C, then the absolute and relative errors are equal to 0.1 °C and 0.0047 °C, respectively. While the absolute error has the same unit as the measurand, the relative error is unitless.

Errors are produced by fluctuations in the output signal and can be *systematic* (e.g., drift or interferences from other systems) or *random* (e.g., random noise).

2.2.7 Noise

The unwanted fluctuations in the output signal of the sensing system, when the measurand is not changing, are referred to as *noise*. The standard deviation value of the noise strength is an important factor in measurements. The mean value of the signal divided by this value gives a good benchmark, as how readily the information can be extracted. As a result, signal-to-noise ratio (S/N) is a commonly used figure in sensing applications. It is defined as:

$$\frac{S}{N} = \frac{\text{Mean value of signal}}{\text{Standard deviation of noise}}. \tag{2.2}$$

Noise can be caused by either *internal* or *external* sources. Electromagnetic signals such as those produced by transmission/reception circuits and power supplies, mechanical vibrations, and ambient temperature changes are all examples of external noise, which can cause systematic error. However, the nature of internal noises is quite different and can be categorized as follows:

1. *Electronic noise*: Thermal energy causes charge carriers to move in random motions, which results in random variations of current and/or voltage. It is unavoidable and is present in all sensing systems operating at temperatures higher than 0 K.

   One of the most commonly seen electronic noises in electronic instruments is caused by the thermal agitation of careers, which is called *thermal noise*.

It produces charge inhomogeneties, which in turn create voltage fluctuations that appear in the output signal. Thermal noise exists even in the absence of current. The magnitude of a thermal noise in a resistance of magnitude R (Ω) is extracted from thermodynamic calculations and is equal to:

$$\bar{v}_{rms} = \sqrt{4kTR\Delta f}, \tag{2.3}$$

in which $\bar{v}_{rms}$ is the root-mean-square of noise voltage, which is generated by the frequency component with the bandwidth of Δf, k is the Boltzmann's constant, which is equal to 1.38×10^{-23} JK^{-1}, and T is the temperature in Kelvin.

Example 1. The rise and fall time of a sensor signal are generally inversely proportional to its bandwidth. Assume that the rise time of a thermistor response is 0.05 s and the relation between the rise time and the bandwidth is $\tau_{rise} = 1/2\Delta f$. (A) Calculate the magnitude of the thermal noise. The ambient temperature is 27 °C and the thermistor resistance is 5 kΩ at this temperature. (B) What is the signal-to-noise ratio, if the average of current passing through the resistor is 0.2 mA?

Answer:

(a) The bandwidth is equal to $\Delta f = 1/2\tau_{rise} = 1/2 \times 0.05$ (s) $= 10$ Hz and according to (2.3), the rms value of the thermal noise voltage is equal to:

$$\bar{v}_{rms} = \sqrt{4 \times 1.38 \times 10^{-23} \times 300 \text{ (K)} \times 5,000 \text{ (Ω)} \times 10 \text{ (Hz)}}$$
$$= 2.88 \times 10^{-8} \text{ (V)} = 0.0288 \text{ mV or } 20 \log(\bar{v}_{rms}) = -150.8 \text{ dB}.$$

(b) Current of 0.2 mA generates a voltage of 5,000 (kΩ) $\times$ 0.0002 (A) $= 1$ V in the thermistor. As a result, the signal-to-noise ratio is:

$$\frac{S}{N} = 1(\text{V})/2.88 \times 10^{-8}(\text{V}) = 3.47 \times 10^9.$$

2. *Shot noise*: The random fluctuations, which are caused by the carriers' random arrival time, produce shot noise. These signal carriers can be electrons, holes, photons, and phonons.

   Shot noise is a random and quantized event, which depends on the transfer of the individual electrons across the junction. Using the statistical calculations, the root-mean-square of the current fluctuation, generated by the shot noise, can be obtained as:

$$\bar{i}_{rms} = \sqrt{2Ie\Delta f}, \tag{2.4}$$

where I is the average current passing through the junction, Δf is the bandwidth, and e is the charge of one electron, which is equal to 1.60×10^{-19} C.

Example 2. In a photodiode the bias current passing through the diode is 0.1 mA. (A) If the rise time of the photodiode is 0.2 ms and the relation between the rise time and the bandwidth is $\tau_{\text{rise}} = 1/4\Delta f$, calculate the rms value of the shot noise current fluctuation. (B) Calculate the magnitude of the shot noise voltage, when the junction resistance is equal to 250 Ω.

Answer: The bandwidth is equal to $\Delta f = 1/4\tau_{\text{rise}} = 1/[4 \times 0.0002 \, (\text{s})] = 1{,}250 \, \text{Hz}$. According to (2.4) the rms value of the shot noise current fluctuation is equal to:

$$i_{\text{rms}} = \sqrt{2 \times 0.1 \times 10^{-3}(\text{A}) \times 1.6 \times 10^{-19}(C) \times 1{,}250 \, (\text{Hz})} = 200 \times 10^{-12}\text{A}$$
$$= 200 \text{ pA}.$$

When the average resistance of the junction is equal to 250 Ω, this fluctuation current generates a rms voltage of $200 \times 10^{-12} \, (\text{A}) \times 250 \, (\Omega) = 50 \times 10^{-9} \, (\text{V}) = 0.05 \, \mu\text{V}$ or $20 \log(\bar{v}_{\text{rms}}) = -146.02 \, \text{dB}$.

3. *Generation-recombination* (or *g-r noise*): This type of noise is produced from the generation and recombination of electrons and holes in semiconductors. They are observed in junction electronic devices.
4. *Pink noise* (or *1/f noise*): In this type of noise the components of the frequency spectrum of the interfering signals are inversely proportional to the frequency. Pink noise is stronger at lower frequencies and each octave carries an equal amount of noise power. The origin of the pink signal is not completely understood.

A term, which is frequently seen in dealing with noise, is *white noise*. White noise has flat power spectral density, which means that the signal contains equal power for any frequency component. An infinite-bandwidth, white noise signal is purely theoretical, as by having power at all frequencies the total power is infinite.

2.2.8   Drift

Drift is observed when a gradual change in the sensing system's output is seen, while the measurand actually remains constant. Drift is the undesired change that is unrelated to the measurand. It is considered a systematic error, which can be attributed to interfering parameters such as mechanical instability and temperature instability, contamination, and the sensor's materials degradation. It is very common to assess the drift with respect to a sensor's *baseline*. Baseline is the output value, when the sensor is not exposed to a stimulus. Logically for a sensor with no drift, the baseline should remain constant.

For instance, in a semiconducting gas sensor, a gradual change of temperature may change the baseline. Additionally, gradual diffusion of the electrode's metal into substrate or sensitive layer may gradually change the conductivity of the sensitive element, which deteriorates the baseline value and causes a drift.

2.2.9   Resolution

Resolution (or *discrimination*) is the minimal change of the measurand that can produce a detectable increment in the output signal. Resolution is strongly limited by any noise in the signal.

A temperature sensing system with four digits has a higher resolution than three digits. When the ambient temperature is 21 °C, the higher resolution system (four digits) output is 21.00 °C while the lower resolution system (three digits) is 21.0 °C. Obviously, the lower resolution system cannot resolve any values between 21.01 °C and 21.03 °C.

2.2.10   Minimum Detectable Signal

In a sensing system, *minimum detectable signal* (*MDS*) is the minimum signal increment that can be observed, when all interfering factors are taken into account. When the increment is assessed from zero, the value is generally referred to as *threshold* or *detection limit*. If the interferences are large relative to the input, it will be difficult to extract a clear signal and a small MDS cannot be obtained.

2.2.11   Calibration Curve

A sensing system has to be calibrated against a known measurand to assure that the sensing results in correct outputs. The relationship between the measured variable (x) and the signal variable generated by the system (y) is called a calibration curve as shown in Fig. 2.2.

2.2.12   Sensitivity

Sensitivity is the ratio of the incremental change in the sensor's output (Δy) to the incremental change of the measurand in input (Δx). The slope of the calibration curve, $y = f(x)$, can be used for the calculation of sensitivity. As can be seen in Fig. 2.2, sensitivity can be altered depending on the calibration curve. In Fig. 2.2, the sensitivity for the lower values of the measurand ($\Delta y_1/\Delta x_1$) is larger than of

Fig. 2.2 Calibration curve:
it can be used for the
calculation of sensitivity

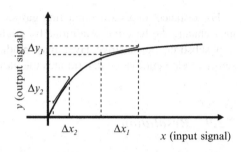

the other section of the curve ($\Delta y_2/\Delta x_2$). An ideal sensor has a large and preferably constant sensitivity in its operating range. An ideal sensor has a large and
preferably constant sensitivity in its operating range. It is also seen that the sensor
eventually reaches *saturation*, a state in which it can no longer respond to any
changes.

For example, in an electronic temperature sensing system, if the output voltage
increases by 1 V, when temperature changes by 0.1 °C, then the sensitivity will be
10 V/ °C.

2.2.13   Linearity

The closeness of the calibration curve to a specified straight line shows the *linearity*
of a sensor. Its degree of resemblance to a straight line describes how linear a
system is.

2.2.14   Selectivity

Selectivity is the sensing system's ability to measure a target measurand in the
presence of others interferences.

For example, an oxygen gas sensor that does not show any response to other gas
species, such as carbon dioxide or nitrogen oxide, is considered a very selective
sensor.

2.2.15   Hysteresis

Hysteresis is the difference between output readings for the same measurand,
depending on the trajectory followed by the sensor.

Hysteresis may cause false and inaccurate readings. Figure 2.3 represents the
relation between output and input of a system with hysteresis. As can be seen,
depending on whether path 1 or 2 is taken, two different outputs, for the same input,
can be displayed by the sensing system.

Fig. 2.3 An example
of a hysteresis curve

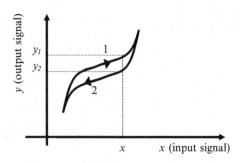

2.2.16 Measurement Range

The maximum and minimum values of the measurand that can be measured with a sensing system are called the *measurement range*, which is also called *the dynamic range* or *span*. This range results in a meaningful and accurate output for the sensing system. All sensing systems are designed to perform over a specified range. Signals outside of this range may be unintelligible, cause unacceptably large inaccuracies, and may even result in irreversible damage to the sensor.

Generally the measurement range of a sensing system is specified on its technical sheet. For instance, if the measurement range of a temperature sensor is between -100 and $800\,°C$, exposing it to temperatures outside this range may cause damage or generate inaccurate readings.

2.2.17 Response and Recovery Time

When a sensing system is exposed to a measurand, the time it requires to reach a stable value is the response time. It is generally expressed as the time at which the output reaches a certain percentage (for instance, 95 %) of its final value, in response to a step change of the input. The "recovery time" is defined conversely.

2.3 Dynamic Characteristics

A sensing system response to a dynamically changing measurand can be quite different from when it is exposed to time invariable measurand. In the presence of a changing measurand, *dynamic characteristics* can be employed to describe the sensing system's transient properties. They can be used for defining how accurately the output signal is employed for the description of a time varying measurand. These characteristics deal with issues such as the rate at which the output changes in response to a measurand alteration and how these changes occur.

Fig. 2.4 Time variation
of a step function

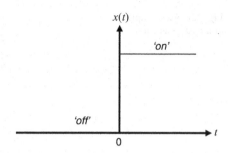

The reason for the presence of dynamic characteristics is the existence of energy-storing elements in a sensing system. They can be produced by electronic elements such as inductance and capacitance, mechanical elements such as vibration paths and mass, and/or thermal elements with heat capacity.

The most common method of assessing the dynamic characteristics is by defining a system's mathematical model and deriving the relationship between the input and output signal. Consequently, such a model can be utilized for analyzing the response to variable input waveforms such as impulse, step, ramp, sinusoidal, and white noise signals.

In modeling a system the initial simplification is always an important step. The simplest and most studied sensing systems are *linear time invariant (LTI)* systems. The properties of such systems do not change in time, hence time invariant, and should satisfy the properties of superposition (addition of two different inputs produces the addition of their individual outputs) and scaling (when input is amplified, the output is also amplified by the same amount), hence linear.

The relationship between the input and output of any LTI sensing system can be described as:

$$a_n \frac{d^n y(t)}{dt^n} + a_{n-1} \frac{d^{n-1} y(t)}{dt^{n-1}} + \cdots + a_1 \frac{dy(t)}{dt} + a_0 y(t)$$
$$= b_m \frac{d^{m-1} x(t)}{dt^{m-1}} + b_{m-1} \frac{d^{m-2} x(t)}{dt^{m-2}} + \cdots + b_2 \frac{dx(t)}{dt} + b_1 x(t) + b_0, \qquad (2.5)$$

where $x(t)$ is the measured (input signal) and $y(t)$ is the output signal and $a_0, \ldots, a_n$, $b_0, \ldots, b_m$ are constants, which are defined by the system's parameters.

$x(t)$ can have different forms such as impulse, step, sinusoidal, and exponential functions. As a simple example, $x(t)$ may be considered to be a step function similar as depicted in Fig. 2.4. This means that a measurand suddenly appears at the sensor. This is an over simplification, as there is generally a rise time when a stimulant appears.

When the input signal is a step change, all derivatives of $x(t)$ with respect to t are zero and (2.5) is reduced to:

$$a_n \frac{d^n y(t)}{dt^n} + a_{n-1} \frac{d^{n-1} y(t)}{dt^{n-1}} + \cdots + a_1 \frac{dy(t)}{dt} a_{n-1} + a_0 y(t) = b_1, \qquad (2.6)$$

for $t \geq 0$ (b_0 is also considered zero in this case. If not zero, a baseline is added to the system response).

Equation (2.6) is a differential equation that models a sensing system response to a step function. Classical solutions to this equation can be readily found in differential equations textbooks and references.

2.3.1   Zero-Order Systems

A perfect *zero-order system* can be considered, if output shows a without-delay response to the input signal. In this case, all a_i coefficients except a_0 are zero. Equation (2.5) can then be simplified to:

$$a_0 y(t) = b_1 \text{ or simply} : \ y(t) = K. \tag{2.7}$$

where $K = b_1/a_0$ is defined as the *static sensitivity* for a linear system.

2.3.2   First-Order Systems

An order of complexity can be introduced when the output approaches its final value gradually. Such a system is called a *first-order system*. A first-order system is mathematically described as:

$$a_1 \frac{dy(t)}{dt} + a_0 y(t) = b_1, \tag{2.8}$$

or after rearranging:

$$\frac{a_1}{a_0} \frac{dy(t)}{dt} + y(t) = \frac{b_1}{a_0}. \tag{2.9}$$

If $\tau = a_1/a_0$ is defined as the time constant, the equation will take the form of a *first-order ordinary differential equation*:

$$\tau \frac{dy(t)}{dt} + y(t) = K. \tag{2.10}$$

This equation can be solved by obtaining the *homogenous* and *particular solutions*.

Solving (2.10) reveals that in response to the step function $x(t)$, $y(t)$ is reaching K via an exponential rate. τ is the time that the output value requires to reach approximately 63% $[(1 - 1/e^{-1}) = 0.6321]$ of its final value K. A typical response for a first-order system is shown in Fig. 2.5.

Fig. 2.5 Response
of a first-order system
to a step function

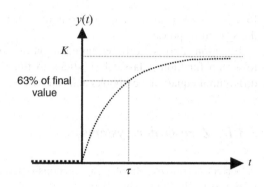

Fig. 2.6 Example 3:
first-order response of a car
temperature sensing system

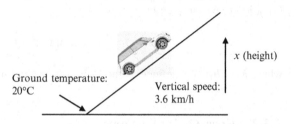

Example 3. A car is equipped with altitude and temperature sensors and associated measurement systems. It is traveling up a hill at a constant speed. This road resembles a ramp (Fig. 2.6) with a constant slope. The temperature at the bottom of the hill is 20 °C. The true temperature at the altitude of x meters is given by: $T_x(x) = 20\,°\mathrm{C} - 0.1x$.

This means that the temperature drops 1 °C for every 10 m of vertical height increase. The altitude measurement system has an ideal (zero-order) response. However, the temperature sensing system has a first-order response with a time constant of $\tau = 10$ s (the delay time).

(a) If the altitude of the car increases with a speed of 3.6 km h^{-1}, what will be the temperature and height measurements at 10, 20, 30, and 40 s?
(b) What will be the values demonstrated by the temperature sensor, if its time constant is reduced to $\tau = 1$ s?

Answer:

(a) The car's altitude is a zero-order function of time t and can be obtained using x (m) = [3,600 (m h^{-1})/3,600 (s h^{-1})] × $t = t$ (s), which means that the altitude increases by 1 m every second. As a result, $T_x(t)$ can also represents the actual ambient temperature as a function of time: $T_x(t) = 20\,°\mathrm{C} - 0.1\,t$.

The measured temperature (the output of a first-order system) is $T_m(t)$, which can be obtained from a first-order differential equation:

$$\tau \frac{\mathrm{d}T_m(t)}{\mathrm{d}t} + T_m(t) = T_x(t).$$

Table 2.1 Temperature sensor responses at different time intervals for the time constant of 10 s

Time (s)	Altitude (m)	Real temp (°C)	Measured temperature (°C)	Temperature error (°C)
0	0	20	20	0
10	10	19	19.6321	0.6321
20	20	18	18.8647	0.8647
30	30	17	17.9502	0.9502
40	40	16	16.9817	0.9817

By substituting the value of the time constant and the function describing the ambient temperature, the resulting equation is:

$$10\frac{dT_m(t)}{dt} + T_m(t) = 20 - 0.1t.$$

A first-order equation's answer is the addition of two general solutions: homogenous (natural response of the equation) and particular integral (generated by the step function in this example). For this example, the homogenous part of the solution is:

$$T_{m-h}(t) = Ae^{\frac{-t}{10}}.$$

The particular-integral part of the solution is given by:

$$T_{m-pi}(t) = -0.1t + 21.$$

Consequently, the total solution can be obtained as:

$$T_m(t) = Ae^{\frac{-t}{10}} + 21 - 0.1t.$$

Applying the initial condition of $T_{mr}(0) = 20$ °C, it can be found that $A = -1$. As a result, the $T_m(t)$ can be calculated by:

$$T_r(t) = -1 \times e^{\frac{-t}{10}} + 21 - 0.1t.$$

Using the above formula, Table 2.1 can be established which presents the values of the ambient temperature (real temperature) and the measured temperature at different intervals.

As can be observed, the error increases with time, approaching 1 °C. This is due to the large time constant value of 10 s. The car has a constant speed, hence constant change of temperature, and there is always a lag in the sensor response. After a while, the system reaches a stable condition where a steady and constant error always exists.

(b) Using a smaller time constant, $\tau = 1$ s, the response of the temperature sensor is faster (a fast processing-responding system) and the differential equation is transformed into:

Table 2.2 Temperature sensor responses at different time intervals for the time constant of 1 s

Time (s)	Altitude (m)	Real temperature (°C)	Measured temperature (°C)	Temperature error (°C)
0	0	20	20	0
10	10	19	19.1	0.1
20	20	18	18.1	0.1
30	30	17	17.1	0.1
40	40	16	16.1	0.1

$$\frac{dT_m(t)}{dt} + T_m(t) = 20 - 0.1t,$$

which has the homogenous solution of:

$$T_{m-h}(t) = Ae^{-t},$$

and the particular-integral solution is given by:

$$T_{m-pi}(t) = -0.1t + 20.1.$$

In this case, the total solution is obtained as:

$$T_m(t) = Ae^{-t} + 20.1 - 0.1t.$$

Using the initial condition of $T_{mr}(0) = 20\,°C$, we obtain $A = -0.1$. As a result, the output temperature equation can be rewritten as:

$$T_m(t) = -0.1 \times e^{-t} + 20.1 - 0.1t.$$

Similarly, Table 2.2 can now be established, which presents the values of the ambient temperature (real temperatures) and the measured temperature at different intervals.

As can be seen for the time constant of 10 s, the error converges to 1 °C but for a time constant an order of magnitude smaller, 1 s, it converges to 0.1 °C.

Example 4. A photodiode sensor has a first-order response with $\tau = 1$ ms. The calibration curve of this sensor is linear and it generates a current of 1 mA at full sun (1 sun) and 0 mA at the fully dark condition. Graph the response of this sensor in time for the following condition: it has been keep in dark for a long time (the initial stable condition), exposed to 0.5 sun for 2 ms, and then turn off the light source to produce the dark condition after that (Fig. 2.7).

Answer: For a first-order system the differential equation is defined as follows:

$$\tau \frac{dy(t)}{dt} + y(t) = K. \tag{2.11}$$

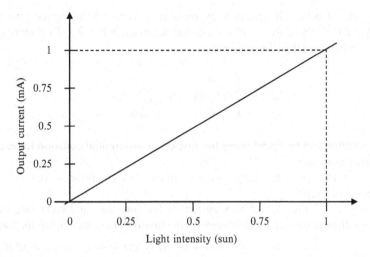

Fig. 2.7 Calibration curve for Example 4

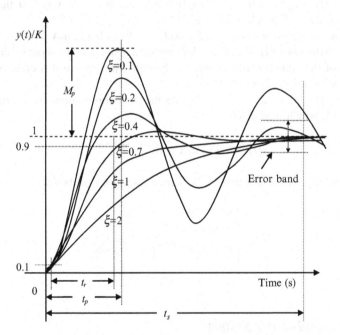

Fig. 2.8 Responses of a second-order sensing system to a step function at different damping ratios

Again, this equation can be solved by obtaining the homogenous and particular solutions.

The value of K can be obtained from the calibration curve as demonstrated in Fig. 2.8.

The sensor has been in zero light exposure ($t < 0$), to 0.5 sun (hence generating 0.5 mA for $0 \leq t \leq 2$ ms), and to zero sun again after $t > 2$ s. Considering the calibration curve, $x(t)$ will be defined as:

$$x(t) = \begin{cases} 0 & t<0 \\ 0.5 \, \text{mA} & 0 \leq t \leq 2\,\text{ms} \\ 0 & t>2\,\text{ms} \end{cases}.$$

The solution can be found using the first-order differential equation for each of these divisors in time.

For $t < 0$ the sensor has long been kept in the dark condition so $y(t) = 0$, as it generates no current.

For $0 \leq t \leq 2$ ms, the 0.5 mA current is fed into the right-hand side of the first-order differential equation. Therefore, the first-order equation takes the format: $(1\,\text{ms}) \frac{dy(t)}{dt} + y(t) = 0.5\,\text{mA}$, solving this equation, the *homogenous* and *particular solutions* will be $y_h(t) = A(e^{-t/1 \ \text{ms}})$ and $y_p(t) = 0.5$ mA, respectively, and: $y(t) = y_h(t) + y_p(t) = A(e^{-t/1 \ \text{ms}}) + 0.5$ mA. As at $t = 0$, $y(t) = 0$ then $A = -0.5$ mA or $y(t) = 0.5$ mA $\times (1 - e^{-t/1 \ \text{ms}})$.

Using this equation $y(2\,\text{ms}) = 0.5\,\text{mA}(1 - e^{-2}) = 0.432$ mA.

For $t > 0$ the sensor is placed in a dark ambient again, and the input on the right-hand side of the equation will eventually be equal to zero. In this case, only the *homogenous* solution exists as: $y(t) = y_h(t) = B \times (e^{-(t - 2\ \text{ms})/1\ \text{ms}})$. As the current is continuous, $y(2\,\text{ms}) = B = 0.432$ mA, which results in:

$$y(t) = 0.432 \ \text{mA} \times (e^{-(t-2\text{ms})/1\text{ms}}).$$

Subsequently, the description of the sensor response is as below:

$$y(t) = \begin{cases} 0 & t<0, \\ 0.5\left(1 - e^{\frac{-t\,\text{ms}}{1\,\text{ms}}}\right) & 0 \leq t \leq 2\text{s}, \\ 0.432\left(e^{\frac{-(t-2\,\text{ms})}{1\,\text{ms}}}\right) & t>2\text{s}. \end{cases}$$

2.3.3   Second-Order Systems

The response of a system can be more complicated. In response to a step function, it may oscillate before reaches its final value. The response can be *overdamped* or *underdamped*. Such responses can be better described by a second-order system approximation.

The response of a *second-order system* to a step change is shown as:

$$a_2 \frac{d^2y(t)}{dt^2} + a_1 \frac{dy(t)}{dt} + a_0 y(t) = b_1. \tag{2.12}$$

By defining the undamped natural frequency as $\omega^2 = a_0/a_2$, and the dampening ratio as $\xi = a_1/2(a_0 a_2)^{1/2}$, (2.12) reduces to:

$$\frac{1}{\omega^2} \frac{d^2y(t)}{dt^2} + \frac{2\xi}{\omega} \frac{dy(t)}{dt} + y(t) = K. \tag{2.13}$$

This is a standard second-order system in response to a step function for which $K = b_1/a_0$.

The damping ratio and natural frequency play pivotal roles in the shape of the response as seen in Fig. 2.8. If $\xi = 0$ there is no damping and the output shows a constant sinusoidal oscillation with a frequency equal to the natural frequency. If ε is relatively small then the damping is light, and the oscillation takes a long time to vanish, *underdamped*. When $\xi = 0.707$ the system is *critically damped*. A critically damped system converges to zero faster than any other conditions without any oscillation. When ξ is large the response is *overdamped*. Other response parameters include: rise time (t_r), peak overshoot (M_p), time to first peak (t_p), and settling time (t_s) (the time elapsed from when the step input is applied to the time at which the amplifier output remains within a specified error band).

Many sensing systems follow the second-order equations. For such systems responses that are not near critically damped condition ($0.6 < \xi < 0.8$) are highly undesirable as they are either slow or oscillatory.

The majority of sensing systems can be nicely described either with the first or second-order equations. However, more complexity can be added when describing a dynamic response of such systems with unusual behaviors. For instance, very often in semiconducting gas sensors, after the initial interactions of the gas with the surface, which is generally a first-order response, many other interactions might occur to change the order of the system. Gas molecules might further diffuse into the bulk of the materials, the morphology of the sensitive material might change, and several stages of interaction might occur. As a result in such systems, obtaining the mathematical description of the dynamic responses can be quite a challenging task.

2.4 Summary

A comprehensive overview of static and dynamic parameters that are used in sensing systems was presented in this chapter.

The most important static characteristics including accuracy, precision, repeatability, reproducibility, stability, error, noise, drift, resolution, MDS, calibration curve, sensitivity, linearity, selectivity, hysteresis, measurement range, as well as response and recovery time were described.

Dynamic characteristics of sensing systems were then presented. Differential equations that describe such systems were discussed with the emphasis of zero-, first-, and second-order systems.

Chapter 3
Physical Transduction Effects

Abstract The most common physical transduction effects are presented in this chapter. They are the effects that are commonly incorporated within the structure of sensors and sensing systems for transforming the target measurands into decipherable signals. Examples of many types of sensors and systems, based on these effects, are presented in each section.

3.1 Introduction

In this chapter some of the major *physical transduction effects*, which are employed in sensor platforms for converting the energy from the measurand into a measurable signal, are presented. For many applications understanding the fundamentals behind these effects is necessary, as this knowledge allows the optimum selection and implementation of the suitably chosen sensors for specific aims.

This chapter will significantly focus on the physical transduction phenomena. The content is a prelude to the major *sensor platforms* that will be presented in the following chapter. Sensor platforms, in contrast to the physical transduction effects, can rely on both chemical and physical phenomena.

3.2 Electromagnetic Effects

Electromagnetic energy is stored and irradiated in the form of *electric charges* and *electromagnetic waves*. When a matter contains an electric charge, it can experience a force when it is placed in the vicinity of another electrically charged matter.

An electric charge is a *quantized physical property*. The smallest quantifiable individual elementary charge is noted by, e, which is approximately 1.602×10^{-19} C. One of the smallest positively charged fundamental elements is proton that has the charge of $+e$, and an electron is one of the smallest negatively charged particles that has the value of $-e$.

K. Kalantar-zadeh, *Sensors: An Introductory Course*,
DOI 10.1007/978-1-4614-5052-8_3, © Springer Science+Business Media New York 2013

The science of charged particles and how they interact with each other is called *electromagnetism*. Electromagnetics (also called *electrodynamics*) is a branch of physics that studies the electromagnetic forces between electric charges. The classical theory of electromagnetism was developed in the nineteenth century by James Clerk Maxwell. The extended version of the classical theory, which investigates the properties of electrical charges at their smallest quantized values, is generally called *quantum electrodynamics* that was developed and expanded during the late nineteenth and early twentieth centuries. This science describes the interaction of electrically charged particles accompanying the exchange of *photons* to obtain a complete picture of matter and electromagnetic field interactions. Photons are basic units of electromagnetic irradiation. Based on quantum electro-dynamics descriptions, it is acceptable to state that electromagnetics science concerns the interaction of photons and incorporated charges in matters, whether they are in the form of atoms, molecules, or larger assemblies.

Electromagnetic radiation consists of mutually perpendicularly propagating electric and magnetic waves, with particle-like properties. The relationships between frequency f, wavelength λ, and energy of the particle E are given by:

$$c = \lambda f, \tag{3.1}$$

$$E = hf, \tag{3.2}$$

where h is the Planck's constant (6.626×10^{-34} J s) and c is the speed of light. In free space, c has the value of 2.998×10^8 m s^{-1}, whereas in other media, its value is adjusted to c/n, in which n is the *refractive index* of the medium.

Maxwell's equations are a set of partial differential equations that describe electromagnetics in the classical forms. These equations are as follows:

$$\nabla \cdot D = \rho_v, \tag{3.3}$$

$$\nabla \times E = -\frac{\partial B}{\partial t}, \tag{3.4}$$

$$\nabla \cdot B = 0, \tag{3.5}$$

$$\nabla \times H = J - \frac{\partial D}{\partial t}, \tag{3.6}$$

where D is the *electric flux density* and E is the *electric field intensity*, which are related by $D = \varepsilon E$, in which ε is the permittivity of the medium. B and H are the *magnetic flux density* and *magnetic field intensity*, respectively, which are related to each other through μ, permeability, as $B = \mu H$. Parameters E, D, B, and H are all vector fields. ρ_v is the *electric charge density per unit volume*, and J is the *current density per unit area*. In the aforementioned equations:

$$\text{Divergence of a vector field } A \text{ is } \nabla \cdot A = \frac{\partial A_x}{\partial x} + \frac{\partial A_y}{\partial y} + \frac{\partial A_z}{\partial z}, \quad (3.7)$$

$$\text{Curl of a vector field } A \text{ is } \nabla \times E = \hat{x}\left(\frac{\partial A_z}{\partial y} - \frac{\partial A_y}{\partial z}\right) + \hat{y}\left(\frac{\partial A_x}{\partial z} - \frac{\partial A_z}{\partial x}\right)$$
$$+ \hat{z}\left(\frac{\partial A_y}{\partial x} - \frac{\partial A_x}{\partial y}\right), \quad (3.8)$$

in which $\hat{x}$, $\hat{y}$, and $\hat{z}$ are the unit vectors in a Cartesian coordinate. Presenting the Maxwell's equations in the integral form, using the divergence and Stoke's theorems, helps in understanding them further:

$$\oint_{\text{Closed surface}} E \cdot dA = \frac{Q_{\text{enc}}}{\varepsilon_v}, \text{ Gauss' law for electricity,} \quad (3.9)$$

$$\oint E \cdot ds = -\frac{\partial \phi_B}{\partial t}, \text{ Faraday – Henry law,} \quad (3.10)$$

$$\oint_{\text{Closed surface}} B \cdot dA = 0, \text{ Gauss' law for magnetism,} \quad (3.11)$$

$$\oint H \cdot ds = i_{\text{enc}} - \varepsilon \frac{\partial \phi_E}{\partial t}, \text{ Ampere – Maxwell law,} \quad (3.12)$$

in which $\oint_{\text{Closed surface}}$ is the integral over all regions over a closed surface, with dA as its infinitesimally small area elements, and $\oint$ specifies the closed loop integral, which means that the calculations are conducted around the loop, with ds as its infinitesimally small length elements. Q_{enc} is the enclosed charge as a result of the area with the charge density ρ_v within the closed surface. ϕ_B and ϕ_E are the magnetic and electric fluxes that are obtained using their associated densities as $\phi_B = \iint_{\text{Area of the closed surface}} B \cdot dA$ and $\phi_E = \iint_{\text{Area of the closed surface}} E \cdot dA$, respectively.

Maxwell's equations describe how electric charges (3.9) and electric currents (3.12) act as sources for the electric and magnetic fields, respectively, and how the electric and magnetic fields are generated from such charges and currents. Equations (3.10) and (3.12) describe how the instantaneous changes in electric and magnetic fluxes crossing a plane alter the magnetic and electric fields in the loops around them, respectively. Equation (3.11) presents that there is no single magnetic field and that magnets always appear in dipoles (in contrast to electric

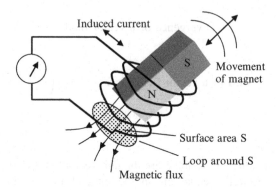

Fig. 3.1 The Faraday–Henry effect. The change of magnetic flux in time induces a current change proportional to this change

charges). Finally, (3.10) and (3.12) describe that magnetic and electric fields coexist.

Very commonly in measuring electromagnetic waves the two associated parameters that are actually measured are current and voltage. So in addition to (3.12), which describes the current density, *electric potential* (or *voltage*), V, which is a scalar parameter, is described as:

$$E = -\nabla V, \tag{3.13}$$

where the gradient of V is defined as:

$$\nabla V = \frac{\partial V}{\partial x}\hat{x} + \frac{\partial V}{\partial y}\hat{y} + \frac{\partial V}{\partial z}\hat{z}. \tag{3.14}$$

The Maxwell's equations, together with (3.13), give us interesting perspective as how electromagnetic waves can be used in sensing applications. According to (3.9) the existence of any charge can produce electric field. Obviously, any transducer that can transform this electric field to a readable signal can be used as a sensor. In many cases, the electric field is transformed into a voltage signal. Transducers such as "capacitors" and "field effect transistors" are two of such examples. If the electric or magnetic fields change in time then according to (3.10) and (3.12) they can produce the corresponding change in fluxes and an associated current according to (3.12) and vice versa. This is the base of many sensors for the detection of electric or magnetic fields for which the current is proportional to the field intensity (transducers such as "inductors").

Faraday–Henry law is one of the fundamental laws of electromagnetism and expresses that an electric field is induced by changing the magnetic field (Fig. 3.1). Michael Faraday and Joseph Henry both independently discovered the electromagnetic phenomenon of self and mutual inductance.

Their work on the magnetically induced currents was the basis of the electrical telegraph, which was jointly invented by Samuel Morse and Charles Wheatstone later on. Early acoustic sensors and devices (such as magnetic microphones), analogue current/voltage meters, and reed-relay switches are all based on this effect. This law is the basis of antennas operation, electrical motors, and an infinite number of electrical devices that include relays and inductors in telecommunication circuits. Almost all *radio frequency identification* (*RFID*) tags and sensing systems, which are currently used in warehouses, are based on the Faraday–Henry effect.

Many affinity sensors operate based on Maxwell's transduction platforms, in which fluxes from magnetic or electric sources cross wires to generate current in them. Obviously, the presence of *permittivity* and *permeability* of the media in the Maxwell's equations also shows that the electromagnetic waves and charges can be used for the measurement of these two fundamental properties of materials.

3.2.1 Electromagnetic Wave Spectrum

Depending on the wavelengths, the type interaction of electromagnetic waves with materials can be used for sensing many of the materials' fundamental properties. When electromagnetic waves interact with materials, the information that is obtained strongly depends on the range of electromagnetic radiation from these materials. From Table 3.1 it is observed that different ranges of incident radiation are responsible for the observation of different phenomena. For example, microwaves stimulate rotations of molecules; infrared radiation stimulates vibrations of molecular vibrational modes; visible and ultraviolet radiation promotes electrons to higher energy orbitals; and X-rays and short wavelengths break chemical bonds and ionize molecules and can even damage soft tissues.

When the near ultraviolet, visible, and near/mid-infrared regions of the electromagnetic spectrum are utilized, the electromagnetic spectroscopy techniques are referred to as *spectrophotometry*. In these techniques, the wavelengths of the incident electromagnetic waves are generally scanned across a range to produce an absorption or emission spectrum. Analogous phenomena occur in the X-rays, microwave, radio, and other regions of the electromagnetic spectrum; however, they will be discussed later in this and subsequent chapters.

Example 1. Calculate the energy of a photon in the X-ray (0.5 nm wavelength) band and for the 630 nm visible region (express them in electron volts unit). In addition to Joules (J), another useful unit of energy, especially in dealing with around the light wavelengths, is electronvolt (1 eV $= 1.602 \times 10^{-19}$ J), which is equal to the kinetic energy gained by an electron, when it accelerates through a 1 V of potential difference.

Answer: The relation between the wavelength and energy of the waves is $E = hf = hc/\lambda$. For the X-ray wavelength of 0.5 nm:

Table 3.1 Types of interactions between electromagnetic radiations of different frequencies with matter

Region	Frequency (Hz)	Wavelength	Example of effects	Energy, E (kJ mol^{-1})
Radio waves	$<3 \times 10^8$	Larger than 1 m	Nuclear and electron spin transitions	$E < 0.001$
Microwaves	3×10^8 to 3×10^{11}	1–10^{-3} m	Molecular rotation	$0.001 < E < 0.12$
Infrared	3×10^{11} to 0.37×10^{15}	10^{-3} m–800 nm	Molecular vibration	$0.12 < E < 150$
Visible	0.37×10^{15} to 0.75×10^{15}	800 nm–400 nm	Electron excitation	$150 < E < 310$
Ultraviolet	0.75×10^{15} to 3×10^{16}	400 nm–10^{-8} m	Electron excitation	$310 < E < 12{,}000$
X-rays	3×10^{15} to 3×10^{19}	10^{-8} to 10^{-11} m	Bond breaking and ionization	$12{,}000 < E < 1.2 \times 10^7$
γ-rays	3×10^{19} to 3×10^{20}	10^{-11} to 10^{-12} m	Nuclear interactions	$1.2 \times 10^7 < E < 1.2 \times 10^8$
Cosmic rays	$>3 \times 10^{20}$	$<10^{-12}$ m		$1.2 \times 10^8 < E$

$$E = \frac{[6.63 \times 10^{-34} \ (\text{J s})] \times [3 \times 10^8 \ (\text{m s}^{-1})]}{0.5 \times 10^{-9} \ (\text{m})} = 3.97 \times 10^{-16} \ (\text{J}).$$

We know that $1 \ (\text{J}) = 1 \ (\text{V})/1.6 \times 10^{-19} \ (\text{C}) = 6.24 \times 10^{18} \ (\text{eV/J})$

$$E = [3.97 \times 10^{-16} \ (\text{J})] \times [6.24 \times 10^{18} \ (\text{eV/J})] = 2.23 \times 10^3 \ (\text{eV}).$$

For the 630 nm wavelength:

$$E = \frac{[6.63 \times 10^{-34} \ (\text{J s})] \times [3 \times 10^8 \ (\text{m s}^{-1})]}{630 \times 10^{-9} \ (\text{m})} = 3.15 \times 10^{-19} \ (\text{J}),$$

$$E = [3.15 \times 10^{-19} \ (\text{J})] \times [6.24 \times 10^{18} \ (\text{eV/J})] = 1.96 \ (\text{eV}).$$

Obviously the energy of the X-ray wave is much larger than the energy of the visible light. This is why our body tissues can be damaged when exposed to X-ray, while the visible light does not cause any short-term harm.

3.3 Dielectric Effect

Materials own unique dielectric properties and are intrinsically made of positive and negative charges. As a result, when they are placed in an electric field, the field can change the relative location of these charges. This produces microscopic dipoles. Considering ε_0 as the permittivity of the free space ($8.85 \times 10^{-12} \ \text{F m}^{-1}$), in the presence of these dipoles the relationship between the *electric flux density*, D, and *electric field intensity*, E, can be written as:

$$D = \varepsilon_0 E + P, \tag{3.15}$$

where P is the *electric polarization field*, which accounts for the polarization property of the materials (Fig. 3.2). A dielectric medium is linear if the magnitude of the induced polarization is proportional to E, and it is isotropic if E and P have the same directions. For such a medium, P is directly proportional to E as:

$$P = \varepsilon_0 \chi_e E, \tag{3.16}$$

where χ_e is the electric susceptibility. Combining (3.15) and (3.16) the permittivity of a material is defined as:

$$\varepsilon = \varepsilon_0 (1 + \chi_e). \tag{3.17}$$

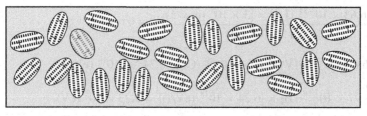

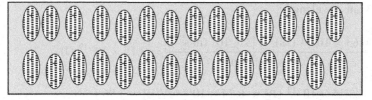

No external
electric field

External
electric field

Fig. 3.2 A dielectric medium polarized by an external electric field

Table 3.2 The relative permittivities of some materials

Material	Relative permittivity, ε_r
Air	1.0006
Polystyrene	~2.6
Glass	4.5–10
Quartz	3.8–5
Mica	5.4–6
PZT (lead zirconate titanate)	>300

This yields the general relationship between E and D as $D = \varepsilon E$. Relative permittivity is then defined as $\varepsilon_r = \varepsilon/\varepsilon_0$. The relative permittivities of some materials are presented in Table 3.2.

Many sensors operate based on the dielectric effects. This includes capacitive and field effect transistor type sensors, which will be explained in Chap. 4. The change in relative permittivity, as a result of exposure to a measurand, is the base of such sensors' operation.

3.4 Permeability Effect

Permeability is the measure of the ability of a material to accommodate magnetic field. The permeability of the free space is denoted by μ_0 and is equal to $4\pi \times 10^{-6}\,\mathrm{H\,m^{-1}}$. The *relative permeability* of a medium is denoted by μ_r and defined as $\mu_r = \mu/\mu_0$ with μ as the permeability of this medium. Similar to the relative permittivity, the relative permeability is also material-dependent parameter and a susceptibility (this time magnetic susceptibility) is defined as $\chi_m = \mu_r - 1$.

Table 3.3 The relative permeabilities of some materials

Material	Relative permeability, μ_r
Permalloy	8,000
Ferrite (manganese zinc)	>650
Steel	100
Aluminum	1.000025
Wood	1.00000045
Air	1
Copper	0.999995
Water	0.999992

Depending on the material, several different properties can be observed during magnetic measurements. The most used ones in sensing transductions are *diamagnetism* and *paramagnetism*. If a material creates a magnetic field in opposition to an externally applied magnetic field, it is a diamagnetic material. Such materials are repulsed by the magnetic field. Diamagnetic materials' relative permeabilities are less than unity. In contrast, paramagnetic materials are attracted to the magnetic field and their relative permeabilities are larger than one. Relative permeabilities of some selected materials are presented in Table 3.3. Obviously such differences in permeabilities allow for the development of magnetic sensors and actuators that take advantage of this property.

3.5 Photoelectric Effect

When a material is irradiated by photons, electrons may be ejected from it. The ejected electrons are called *photoelectrons*, and their kinetic energy, E_K, is equal to the incident photon's energy, hf, minus some threshold energy, known as the *material's work function* ϕ, which needs to be exceeded for the material to release electrons. The effect is illustrated in Fig. 3.3 and is governed by:

$$E_K = hf - \varphi, \tag{3.18}$$

where h is the Planck's constant ($h = 6.625 \times 10^{-34}$ J s) and f is the photon's frequency.

In the past, the photoelectric effect has been traditionally used in vacuum tube amplifiers. This effect can also be used for developing special types of electromagnetic sensors. Because the work function depends on the material, sensors may be designed that are tuned to specific wavelengths. Such sensors are widely used for photoelectron microscopy. In such devices, the work function of a material is obtained by bombarding it with a monochromatic X-ray or UV source and measuring the kinetic energy of the emitted electrons.

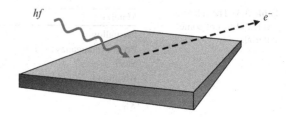

Fig. 3.3 Photoelectric effect:
incident of a photon
and release of an electron
as a result

3.6 Photoconductive Effect

Photoconductivity occurs when a beam of photons impinges a semiconducting
material, causing its conductivity to change. The incident photons excite electrons
from the conduction band to the valence band, which occurs if the light striking the
semiconductor has sufficient energy (hf is the energy of the incident photon that
has the frequency of f). Light response depends on the *bandgap* of the materials.
A simple depiction of electronic band structures for two semiconductors with
different bandgaps is shown in Fig. 3.4. Obviously for a wider bandgap material
larger energy is required to excite an electron from the valence band to the
conduction band.

The photoconductive effect is widely utilized in electromagnetic radiation
sensors, and such devices are termed *photoconductors, light-dependent resistors*
(LDR), or *photoresistors. Cadmium sulfide* (CdS—bandgap of ~2.42 eV, which is
approximately the wavelength of 512 nm in the violet region) and *cadmium*
selenide (CdSe— ~1.73 eV, which is approximately the wavelength of 716 nm in
the yellow region) are the current materials of choice for the fabrication of photoconductive devices and sensors. Very commonly to make a photoresistor, a film of
these materials is deposited onto parallel electrodes (Fig. 3.5). Devices based on
semiconductors such as CdS can have a wide range of resistance values, from about
a few ohms, when the light has high intensity, to several mega ohms in darkness.
They are capable of responding to a broad range of photon wavelengths, including
infrared, visible, and ultraviolet. The response time of such semiconducting films
ranges between milliseconds and tens of seconds, depending on the porosity,
thickness, and other physical properties of the films.

3.7 Photovoltaic Effect

In the photovoltaic effect, a voltage is induced by absorbed photons at a junction of
two dissimilar materials (it is also called a *heterojunction*). The absorbed photons
produce free charge carriers (electrons and holes) as can be seen in Fig. 3.6.
However, the induced voltage in the heterojunction causes the charge carriers to
move apart, resulting in current flow in an external circuit. Materials used for

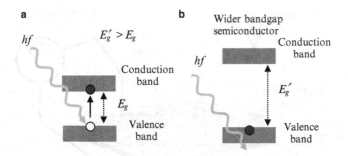

Fig. 3.4 Depiction of the electronic structures for two semiconductors with different bandgaps: (a) incident photon energy is larger than the bandgap ($hf > E_g$), so it excites an electron from the valence band to the conduction band, and (b) incident photon energy is smaller than the bandgap energy ($hf < E_g'$), so it does not affect the semiconductor

Fig. 3.5 Photo of a commercial LDR based on CdS

fabricating such heterojunctions are generally semiconductors, which are responsive to light of various wavelengths.

A typical photovoltaic device is seen in Fig. 3.6. They generally consist of a large area semiconductor p–n junction or diode. A photon impinging on the junction is absorbed, if its energy is greater than or equal to the semiconductor's bandgap energy. This can cause a valance band electron to be excited into the conduction band, leaving behind a hole, and thus creating a mobile electron–hole pair. If the electron–hole pair is located within the depletion region of the p–n junction, then the existing electric field will either sweep the electron to the n-type side or the hole to the p-type side. As a result, a current is generated as below:

$$I = I_S[e^{\frac{qV}{kT}} - 1], \tag{3.19}$$

where q is the electron charge (1.602×10^{-19} C), k is the Boltzmann's constant (1.38×10^{-23} J/K^{-1}), and T is the temperature of the p–n junction in Kelvin.

Photovoltaic cells and sensors are commonly made from materials that absorb photons in the infrared, visible, and UV ranges: materials such as silicon (wavelengths between 190 and 1,100 nm), germanium (800–1,700 nm), indium gallium arsenide (800–2,600 nm), and lead sulfide (1,000–3,500 nm) are generally used.

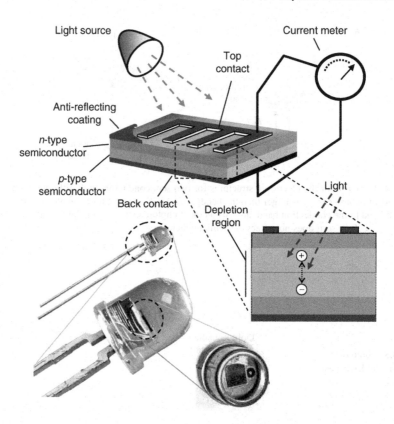

Fig. 3.6 Schematic of a photovoltaic device and the photo of a typical photovoltaic sensor

The most common photovoltaic-based sensors are photo transistors and photo diodes which will be further explained in Chap. 4. Photovoltaic devices can be employed in a wide range of sensing applications. These include use in analytical apparatus such as spectrophotometers, radiation monitors, automatic light adjustment systems in buildings, as light sensors in optical communication systems, detector in vehicles and lifts, in electronic appliances, in cookware and industrial surveillance systems. Photovoltaic devices are also the basis of photovoltaic cells.

3.8   Photodielectric Effect

Materials, whose dielectric properties change when illuminated by electromagnetic radiation, are called photodielectric materials. In such materials, permittivity is changed upon exposure to the radiation, as a result of the atoms or molecules' transitions to excited states in which polarizability differs from that of in the ground state.

Photodielectric measurements have been widely employed in photochemistry as well as to study kinetics of processes in photographic materials and semi-conductors. Compounds of semiconductors with photodielectric effect are commonly used for the detection of low intensity electromagnetic radiations.

3.9 Photoluminescence Effect

In *photoluminescence* (commonly abbreviated as *PL*), light is emitted from atoms or molecules after they have absorbed photons. The absorbed photons give their energy to the material, causing it to change to an excited energy state. Then after some time, the material radiates the excess energy back out in the form of a photon, and consequently returns to a lower energy state. The energy of the emitted light relates to the difference in energy levels of the excited state and the equilibrium state. *Fluorescence* and *phosphorescence* are examples of photoluminescence.

3.9.1 Fluorescence and Phosphorescence

Photoluminescence can be understood through quantum mechanics. It depends on the *electronic structure* of atoms and molecules. Molecules have electronic states, and within each there are different *vibrational levels* (Fig. 3.7). After accepting energy in the form of a photon, an electron is raised to an excited electronic state. For most molecules, the electronic states can be divided into *singlet* (S) and *triplet* (T) *states*, depending on the electron spin. After a molecule is excited to a higher electronic energy state, it loses its energy quite rapidly via a number of pathways (see Fig. 3.7).

In fluorescence, vibrational relaxation brings the molecule to its lowest vibrational energy level, $V' = 1$, in the first excited singlet state, S_1. Consequently, the electron relaxes from the lowest vibrational energy level in S_1 to any vibrational level of S_0. For phosphorescence, the electron in S_1 undergoes *intersystem crossing* to T_1 and then relaxes to S_0. Due to the multiple rearrangements during the process, the phosphorescence has a much longer lifetime than the fluorescence. For fluorescence, the period between absorption and emission is typically between 10^{-8} and 10^{-4} s. However for phosphorescence, this time is generally longer (10^{-4} to 10^2 s).

Nowadays, many analytical measurements involve the use of fluorescence materials. An ideal fluorescence material for sensing usually has a single exciting wavelength and single detection wavelength. Microscopes or spectrometers are then used for detecting the presence of the signal at the desired wavelengths. The sensitivity of the method can be quite high and even single fluorescent molecule concentration can be detected.

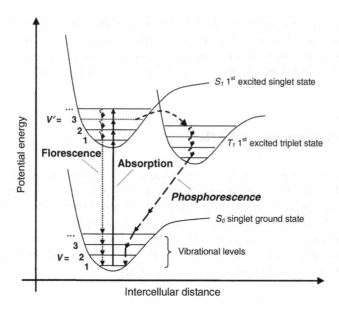

Fig. 3.7 A depiction of fluorescence and phosphorescence processes

Organic molecules that exhibit fluorescence find many applications in sensing. For example, fluorescence probes are used in biotechnology as tools for monitoring biological events in individual cells. Fluorescence microscopy of tissues, cells, or subcellular structures can be carried out by labeling an antibody with a *fluorophore*. Labeling multiple antibodies with different fluorophores allows the visualization of multiple targets within a single image. Many molecules are used as *ion probes*, as after interacting with ions (e.g., Ca^{2+}, Na^+, etc.) they fluoresce. Such ion probes are important in neurological processes, within which their photoluminescence properties such as absorption wavelength, emission wavelength, or emission intensity may change. Another important application of such ion probes is for sensing *heavy metal ions*. As an example Fig. 3.8 demonstrates response of a type of fluorescence probe (boradiazaindacene fluorophore) which is selective to cadmium ions (Cd^{2+}) [1]. Cd^{2+} is toxic to bones, kidneys, nerve systems, and tissues, may cause renal dysfunction, metabolism disorders, and increase cancer incidence. Since Cd^{2+} can accumulate in organisms, there is a great need for sensors that can monitor its levels in living cells or tissues.

Zinc sulfide (ZnS) and *strontium aluminate* ($SrAl_2O_4$) are two of the most common phosphorescent materials. Such materials have been widely used for safety related products and sensors. However, as the luminance of $SrAl_2O_4$ is approximately an order of magnitude greater than ZnS, it is now used in most of the phosphorescence-related applications. $SrAl_2O_4$ is regularly used in applications such as in pathway marking and other safety related signages.

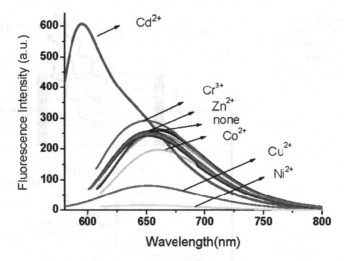

Fig. 3.8 The fluorescence spectra of a sensor incorporating boradiazaindacene (5 μM) as the fluorophore in the presence of different metal ions (150 μM) in HCl (0.01 M) solution, (reproduced from [1] with permission)

3.9.2 Electroluminescence

Electroluminescence occurs when a material emits light as a result of an electrical current flowing through it or when subjected to an electrical potential. It is the conversion of electrical energy into radiant energy. There are two main ways of producing electroluminescence: by passing a current through (1) a heterojunction or (2) a phosphorescent material.

Electroluminescence can occur when a current passes through a heterojunction (such as at the junction of p–n doped semiconducting materials). Electrons can recombine with holes, causing them to fall into a lower energy level and release energy in the form of photons. Such a device is commonly called a *light-emitting diode* (*LED*) . The layout of a typical LED is shown in Fig. 3.9.

The wavelength of the emitted light is determined by the bandgap energy of the materials forming the junction. However, the flow of a current does not always guarantee electroluminescence. In diodes made of indirect bandgap materials, such as silicon, the recombination of electrons and holes is non-irradiative and there is no light emission. Materials used in LEDs must have a direct bandgap. Those comprised of periodic table III and V element compounds, which are most commonly used in the fabrication of LEDs. These include III–V semiconductors such as *GaAs* and *GaP*. It is important to mention that the bandgap of these materials, and hence the emission wavelength, can be tailored by adding impurities. For instance, LEDs made solely from GaP emit green light at 555 nm. However, nitrogen-doped GaP emits at yellow-green light (565 nm), and ZnO-doped GaP emits red light (700 nm).

Fig. 3.9 Schematic of an LED. The band diagram illustrates the electron–hole recombination process and the subsequent emission of photons

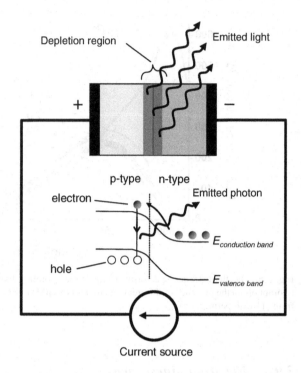

Fig. 3.10 Schematic of an electroluminescence device based on a phosphorous material

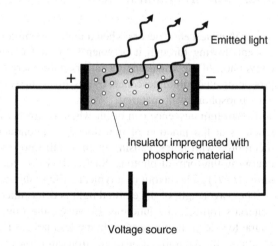

The other way in which electroluminescence can occur is via the excitation of electrons using an electric field that is applied across a phosphorescent material. This type electroluminescence stems from the work of Georges Destriau, who in 1936 showed that by applying a large alternating potential across zinc sulfide phosphorous powder suspended in an insulating material, electroluminescence would be exhibited (Fig. 3.10).

Electroluminescent devices are implemented in the integrated sensors for spectroscopy equipment. Also, many disposable sensors use the electroluminescence effect. Such sensors operate based on assessment of the illuminated light intensity as the measure of a measurand quantity. For instances, many of the pregnancy testing systems are equipped with such LED which are illuminated upon the detection of positive outcomes. The electroluminescent effect is also an integrated part of many electrochemical sensing systems (electrochemical sensing templates will be presented in Chap. 4). When an electron is generated in an electrochemical interaction, it can transform into a photon via the usage of electroluminescent devices. Consequently, this irradiation can be detected with a photodiode or photo transistor. Advantageously, the use of optical reading reduces the electronic noise and also it is compatible with many standard optical sensing systems.

Electroluminescent devices have many applications in chemical sensing. The change in the electroluminescence of metal oxides such as TiO_2 and ZnO can be utilized for measuring the concentration of oxidizing materials such as hydrogen peroxide. In such systems, the analyte molecules lose their oxygen on the metal oxide's surface (hence releasing an electron into the film), resulting in electroluminescence intensity change proportional to the concentration of measurand.

3.10 Hall Effect

Discovered in the 1880s by Edwin Hall, when a magnetic field is applied perpendicularly to the direction of an electrical current flowing in a conductor or semiconductor, an electric field arises that is perpendicular to both the direction of the current and the magnetic field. As can be seen in Fig. 3.11, a magnetic field is applied perpendicularly to a thin sheet of material that is carrying a current. The magnetic field exerts a transverse force, F_B, on the moving charges and pushes them to one side. These results in electric charges to build up on one side and regions of positive and negative charges are formed. This charge separation creates an electric field that generates an electric force, F_E. Eventually, this electric force balances the magnetic force by applying an opposing force to the charges. As a result of the charge separation, a measurable voltage between the two sides of the material, called the Hall voltage, V_{Hall}, is established, which is obtained using:

$$V_{Hall} = \frac{IB}{ned}, \tag{3.20}$$

where I is the current flowing through the material, B is the magnetic field, n is the charge carrier density of the material, e is the electron charge equal to 1.602×10^{-19} C, and d is the thickness of the material.

Hall effect is one of the most widely used effects in sensor technology, particularly for monitoring magnetic fields. Commercial Hall effect sensors are utilized in sensing of fluid flow, current, power, and pressure as well as the measurement of magnetic fields.

Fig. 3.11 The schematic of
a Hall effect device (before
and after applying a magnetic
field, B)

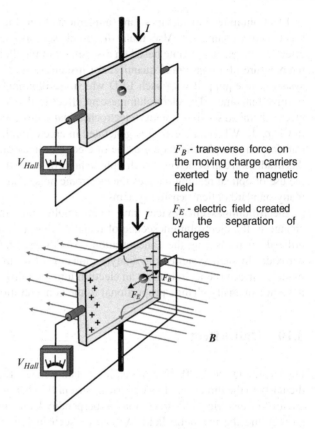

F_B - transverse force on
the moving charge carriers
exerted by the magnetic
field

F_E - electric field created
by the separation of
charges

Example 2. A large current of 100 A passes through a slab of copper with the
thickness of 2 mm and produces a voltage of 1 μV. The slab is exposed to a
magnetic field of 0.2 T. From this measurement obtain the concentration of
electrons in copper.

Answer:

$$n = \frac{100 \ (\text{A}) \times 0.2 \ (\text{T})}{1 \times 10^{-6} \ (\text{V}) \times 1.6 \times 10^{-19} \ (\text{C}) \times 2 \times 10^{-3} \ (\text{m})} = \frac{20}{3.2 \times 10^{-28}}$$
$$= 6.25 \times 10^{28} \ (\text{m}^{-3}) = 6.25 \times 10^{22} \ (\text{cm}^{-3})$$

This number is actually quite close to the real value of 8.5×10^{22} cm^{-3} for
copper.

Example 3. Use a silicon-based Hall effect sensor with the substrate thickness of
250 μm and the electron concentration of 10^{18} cm^{-3}. What kind of voltage is
produced in this sensor, if a magnetic field of 2 mT is applied? The constant current
that passes through this device is 10 μA.

Answer:

$$V_{\text{Hall}} = \frac{0.002 \ (\text{T}) \times 10^{-5} \ (\text{A})}{10^{15} \ (\text{cm}^{-3}) \times 1.6 \times 10^{-19} \ (\text{C}) \times 250 \times 10^{-6} \ (\text{m})} = 5 \ \text{mV}.$$

3.11 Thermoelectric (Seebeck/Peltier and Thomson) Effect

The thermoelectric effect is the direct conversion of temperature differences to electric voltage and vice versa. It was first observed in early 1800s by Thomas Johann Seebeck and Jean Charles Athanase Peltier separately; hence also called *Seebeck–Peltier effect*.

As seen in Fig. 3.12, for two dissimilar materials, A and B, a voltage difference ΔV is generated when two junctions are held at different temperatures. The voltage difference is proportional to the temperature difference, $\Delta T = T_2 - T_1$, and the relationship is given by:

$$\Delta V = (S_A - S_B)\Delta T, \tag{3.21}$$

where S_A and S_B are the Seebeck coefficients of materials A and B, respectively. This phenomenon provides the physical basis for *thermocouples*, one of the standard devices for measuring temperature.

The exact opposite observation occurs when a temperature difference arises at a junction of two dissimilar metals, while a current is passing through them (Fig. 3.12). The heat per unit time, Q, absorbed by the lower temperature junction is equal to:

$$Q = (\Pi_A - \Pi_B)I, \tag{3.22}$$

where Π_A and Π_B are the Peltier coefficients of each material and I is the current. Depending on the current magnitude heat leaves or accumulated in the junction.

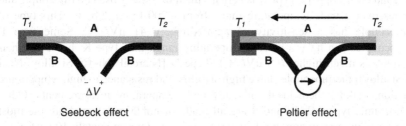

Seebeck effect   Peltier effect

Fig. 3.12 Two dissimilar materials A and B in intimate contact, with either ends are held at different temperatures (T_1 and T_2)

Table 3.4 Some common types of thermocouples

Type	Materials	Temperature range (°C)
K	Chromel/alumel (Ni–Al alloy)	−200 to +1,200
E	Chromel/constantan	−110 to 140
J	Iron/constantan	−40 to +750
N	Nicrosil (Ni–Cr–Si alloy)/Nisil (Ni–Si alloy)	

A complementary effect was also discovered by William Thomson (Lord Kelvin) in 1850s. He found that an electric current flowing along a material, which experiences a temperature gradient along its length, can cause the material to either absorb or release heat per unit volume. The *Thomson effect* is described by the following equation:

$$Q = \rho J^2 - \mu J \frac{dT}{dx}, \tag{3.23}$$

where J is the current density passing though the materials, ρ is the resistivity of the material, dT/dx is the temperature gradient along the material, and μ is the Thomson coefficient. There are two terms on the right-hand side of the equation: (a) the first term is the Joule heating and (b) the second term is the Thomson heating.

Thomson, Seebeck, and Peltier coefficients are related. If the Thomson coefficient is integrated over a wide temperature range, it can be integrated using (3.23) to obtain the absolute values for the Seebeck and Peltier coefficients. Of course, if the value of Seebeck coefficient of a standard material is known, other materials can be measured against such a reference in junctions. The most commonly used reference material is one that has the Thomson coefficient of zero.

Many sensing systems incorporate temperature sensors based on the thermoelectric effect, and there are a variety of them available that find application in medical and scientific research, as well as in industrial process control and food storage systems. There are several types of such devices, called thermocouples, and among the most popular are listed in Table 3.4.

The different metals and alloys utilized in thermocouples result in different properties and performance. Some commonly utilized alloys are chromel (approximately 90 % nickel and 10 % chromium) and constantan (approximately 40 % nickel and 60 % copper). Type K is perhaps the most widely used thermocouple and operates over a wide temperature range from −200 to +1,200 °C. This type of thermocouple has a sensitivity of approximately 41 μV/ °C. Some type E thermocouples can have a narrower operating range than type K, however, their sensitivity is much higher (68 μV/ °C). Type N [Nicrosil (Ni–Cr–Si alloy)/Nisil (Ni–Si alloy)] thermocouples have high stability and resistance to high temperature oxidation, making them ideal for many high temperature measurements. Other thermocouple types: B, R, and S are all made of noble metals and are the most stable for high temperature, but have low sensitivity (approximately 10 μV/ °C).

Thermoelectric materials, particularly those based on semiconducting materials with large Peltier coefficients, can be used to fabricate on chip temperature sensors.

They are also used to make heat pumps, as nowadays, many products including *charge coupled device* (*CCD*) cameras, laser diodes, microprocessors, and blood analyzers employ thermoelectric coolers.

The performance of thermoelectric devices in terms of the ability to convert thermal energy into electrical energy, and vice versa, depends on the *figure of merit* (*ZT*) of the material's utilized and is given by:

$$ZT = (S^2T)/(\rho K_T), \tag{3.24}$$

where S, T, ρ, and K_T are the Seebeck coefficient, absolute temperature, electrical resistivity, and total thermal conductivity, respectively. Generally, the larger the thermoelectric material's Seebeck coefficient (to generate the maximum voltage difference) and thermal conductivity (so it does not allow the exchange of heat at two junctions), whilst the lower its electrical resistivity (so the internal resistance does not generate heat), the more efficient the thermoelectric devices can be made.

In the 1950s, an Australian researcher, Julian Goldsmid confirmed that *bismuth telluride* (Bi_2Te_3) and *antimony telluride* (Sb_2Te_3) display very strong Peltier effects. Bi_2Te_3 and Sb_2Te_3 are semiconductor materials with high Seebeck coefficients (~200–250 μV/ °C) with *ZT* of approximately equal to unity at room temperature.

Example 4. Seebeck coefficients of nickel and tungsten are -15 and 7.5 μV/ °C, respectively. We make thermocouple is made of the junction of these two metals. What will be the voltage shown at 100 °C, if the voltage generated at 23 °C is zero?

Answer: The temperature difference is $100 - 23$ °C $= 77$ °C and the difference of the Seebeck coefficients is 22.5 μV/ °C. As a result, the voltage generated is:

$$\Delta V = (77(^\circ C)) \times (22.5 \text{ (mV per } ^\circ C)) = 1,732 \text{ (mV)}.$$

3.12 Thermoresistive Effect

Thermoresistive effect occurs as a result of the change in a material's electrical resistance upon temperature change and is widely used in temperature sensing applications. This effect is the basis of temperature sensing devices such as *resistance thermometers* and *thermistors*. The electrical resistance, R, is determined by the formula:

$$R = R_{\text{ref}}\left(1 + \alpha_1\Delta T + \alpha_2\Delta T^2 + \cdots + \alpha_n\Delta T^n\right), \tag{3.25}$$

where R_{ref} is the resistance at the reference temperature, $\alpha_1 \ldots \alpha_n$ are the material's temperature coefficient of resistance, $\Delta T = (T - T_{ref})$ is the difference between the current temperature T and the reference temperature T_{ref}. The resistance can either increase or decrease with the temperature change. If the material has a *positive temperature coefficient* (*PTC*) then its resistance increases with temperature, whereas if it has a *negative temperature coefficient* (*NTC*), then it decreases. If the thermoresist materials exhibit near linear relationships between the temperature and resistance, then the higher order terms in (3.25) can be disregarded. However, the linearity is generally valid only for a limited range of temperatures.

Thermistors are commonly made of ceramics or polymers. Such thermistors generally operate within a limited temperature range in the order of −50 to 150 °C. Thermoresistive effect is also observed in metals. Such elements are typically referred to as *resistance temperature detectors* (*RTD*), which are implemented over larger temperature ranges.

Example 5. If first-order temperature coefficient of resistance of a thermoresistive material (α_1) is equal to 30 Ω/ °C, how much change in resistance is measured as a result of a 20 °C change of temperature?

Answer:

$$\Delta R = [30 \ (\Omega/^\circ C)] \times [20 \ (^\circ C)] = 600 \ (\Omega).$$

3.13 Piezoresistive Effect

The *piezoresistive effect* describes the change in a material's electrical resistivity, when a mechanical force is exerted on it. This effect took its name from the Greek word *piezein*, which means to squeeze. It was first discovered in 1856 by Lord Kelvin, who found that the resistivity of certain metals changed upon applying a mechanical load. Semiconductors such as silicon and germanium typically show large piezoresistive effects. There are two phenomena attributed to a semiconductor's resistivity change: the stress-dependent changes of its geometry and the stress dependence of its resistivity. For these semiconductors, the first-order piezoresistance can be described using:

$$\frac{\Delta R}{R} \approx \pi\sigma, \tag{3.26}$$

where π is the first-order *tensor element of the piezoresistive coefficient*, σ is the mechanical stress tensor, and R and ΔR are the resistance and the change in resistance, respectively.

The piezoresistive effect in semiconductors and metal alloys is used in sensors. Most materials exhibit some piezoresistive effect. However, as silicon is the material of choice for integrated circuits, the use of piezoresistive silicon devices, for mechanical stress measurements, has been of great interest. Many commercial devices such as pressure sensors and acceleration sensors employ the piezoresistive effect in silicon.

The piezoresistive effect also frequently used in cantilever-based sensors, which are implemented in low-dimensional sensing systems together with atomic force microcopy equipment or as microscales both in liquid and gaseous media.

3.14 Piezoelectric Effect

Piezoelectric effect is the ability of crystals that lack a center of symmetry to produce a voltage in response to an applied mechanical force and vice versa (Fig. 3.13). The effect was discovered by the Curie brothers in 1880. The nature of the effect is related to the occurrence of electric dipole moments. In a piezoelectric material the polarization can alter when a mechanical stress is applied. When stress is exerted, the change of polarization might either be caused by a re-configuration of the dipole-inducing surrounding or by re-orientation of molecular dipole moments. The piezoelectric effect depends on: (1) the orientation of polarization, (2) crystal symmetry, and (3) the applied mechanical stress.

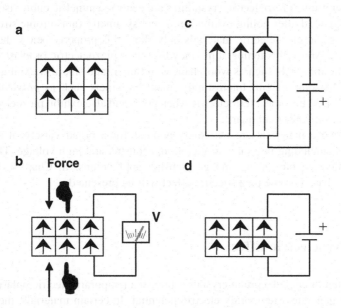

Fig. 3.13 (a) A piezoelectric material. (b) A voltage response can be measured as a result of a compression or expansion. An applied voltage (c) expands or (d) compresses a piezoelectric material depending on its polarity

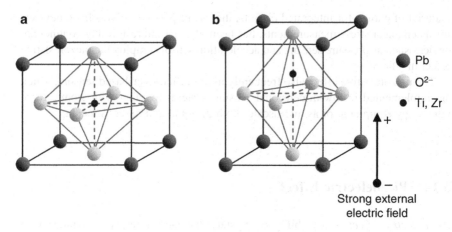

Fig. 3.14 Piezoelectricity in a PZT crystal: (**a**) before poling; (**b**) after poling

Out of 32 crystal classes, 21 do not have a center of symmetry (non-centro-symmetric), and of these, 20 directly exhibit piezoelectricity (except the cubic class 432). The most popular piezoelectric materials are quartz, *lithium niobate*, *lithium tantalite*, *lead zirconate titanate (PZT)*, *barium titanate* (BaTiO$_3$), *lead titanate* (PbTiO$_3$), ZnO, and lanthanum gallium silicate (langasite). Many piezoelectric materials are ceramics, which become piezoelectric when poled with a strong eternal electric field (Fig. 3.14). The poling process should generally occur at high temperatures. Piezoelectric crystallites are centro-symmetric cubic (isotropic) before poling and after poling exhibit tetragonal symmetry (anisotropic structure). After poling the piezoelectric materials only show this property below the Curie temperature. Above this temperature, they lose their piezoelectric properties.

Inorganic materials such as wool, hair, wood fiber, and silk also exhibit piezo-electricity to some extent. Interestingly, many polymers such as *polyvinylidene fluoride (PVDF)* become a ferroelectric when poles and also exhibit piezoelectricity several times greater than quartz.

Piezoelectric materials are extremely popular for a broad variety of sensing applications including pressure, acceleration, acoustic, and high voltage. They are also prevalent in biosensing and gas sensing applications. In Chap. 4, several transducers based on the piezoelectric effect will be presented.

3.15   Pyroelectric Effect

When heated or cooled, certain crystals establish a temporary electric polarization, and hence generate a temporary electric potential. In certain materials, the temperature change causes positive and negative charges to be displaced to the opposite ends of a crystal's polar axis. Such polar crystals are said to exhibit

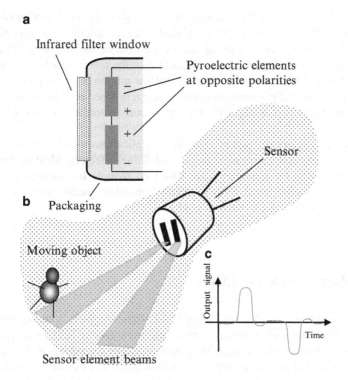

Fig. 3.15 (a) Schematic of a pyroelectric sensor based on the elements in one package, (b) passage of an object that exudes infrared in front of the sensor, and (c) the output signal

pyroelectric effect, which takes its name from the Greek word *pyro* which means fire. If the temperature stays constant at its new value, the pyroelectric voltage gradually disappears due to leakage current. The most common pyroelectric materials include *lithium tantalate*, *gallium nitride* (*GaN*), *cesium nitrate* (*CsNO₃*), *polyvinyl fluorides*, and *PZT*.

The pyroelectric materials are generally employed in radiation sensors, in which radiation incident on their surface is converted to heat. The increase in temperature associated with this incident radiation causes a change in the magnitude of the material's polarization. This results in a measurable voltage, or if placed in a circuit, a measurable current is given by:

$$I = pA\frac{dT}{dt},$$ (3.27)

where p is the pyroelectric coefficient, A is the area of the electrode, and dT/dt is the temperature rate of change.

Pyroelectric materials are often used as infrared and millimeter wavelength elements. Irradiation sensors based on the pyroelectric effect are now commercially available. Common pyroelectric infrared motion sensors have two sensing elements, which are internally connected in a voltage bucking configuration as shown in Fig. 3.15. The two sensors are placed in reverse polarities; as a result when an infrared emitting body is moving across the front of the sensor, it affects the adjacent pyroelectric element, then both, and then the other pyroelectric element. The output signal from the sensor shows that for motion in one direction, first a positive, then zero, and eventually a negative signal is observed. Motion in the other direction produces first a negative, then zero, and eventually a positive output signal.

Pyroelectric sensors are used in a variety of applications such as motion sensors, light control, temperature measurements, and flame detectors. Most of the measurement standards for radiometry are based on pyroelectric thermal detectors. These devices employ some form of thermal-absorber coating such as carbon-based paint or diffuse metals such as gold.

3.16   Magnetostriction Effect

Magnetostriction, also called the *magneto-mechanical effect*, is the change in a material's dimensions when subjected to an applied magnetic field, or alternatively it is a change in a material's magnetic properties under the influence of stress and strain. It was first identified in 1840s by James Joule, while examining a sample of nickel. Its name originates from the Greek word, *magnet* and the Latin word *strictus* (meaning compressed, pressured, tense).

Macroscopically there are two distinct mechanisms that occur in magnetostriction, which are illustrated in Fig. 3.16. These are: the rotation of the magnetic domains (arrows all pointing in the same direction) and the migration of domain walls within the material (i.e., the domains lined up end-to-end). The change in the material's dimension in response to the externally applied field is a culmination of these two mechanisms.

Magnetostrictive materials convert magnetic energy into kinetic energy and vice versa. Therefore, they are regularly used for sensing and actuation upon exposure to

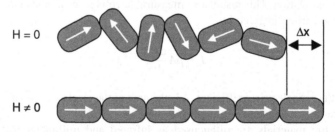

Fig. 3.16 Magnetostriction: unaligned magnetic domains (*top*) will align causing the structure to expand under the influence of an applied magnetic field (*bottom*)

magnetic fields. Interestingly, this effect is the causes of the familiar humming sound that is heard in electrical transformers.

Magnetostriction is defined by the magnetostrictive coefficient, Λ. It is defined as the fractional change in length as the magnetization of the material increases from zero to the saturation value. The coefficient, which is typically in the order of 10^{-5} (or 10 microstrains), can be either positive or negative. This value is 60 microstrains for cobalt which is the largest room temperature magnetostriction of any pure element. The reciprocal of this effect is called the *Villari effect*, where a material's susceptibility changes when subjected to a mechanical stress.

The most advanced magnetostrictive materials, called *Giant Magnetostrictive* (*GM*) materials, are alloys composed of iron (Fe), dysprosium (Dy), and terbium (Tb). Many of which were discovered at Naval Ordnance Lab and Ames Laboratory in mid-1960s [2]. Their Λ values can be one or two orders of magnitude larger than those of pure elements. The GM effect is commonly used for the development of magnetic field, current, proximity, and stress sensors.

3.17 Magnetoresistance Effect

Magnetoresistance is the dependence of a material's electrical resistance on an externally applied magnetic field and it was first observed by Lord Kelvin in 1850s. The applied magnetic field causes a Lorentz force to act on the moving charge carriers, and depending on the field's orientation, it may result in a resistance to the flow of current. For an electric charge, in an electric field intensity of E, the velocity is calculated as $v = \mu E$. If a magnetic flux density of B is also present, the velocity then satisfies the following equation:

$$v = \mu(E + v \times B), \tag{3.28}$$

in which μ is the carrier mobility. The velocity can then be extracted as:

$$v = \mu\left(\frac{E + \mu E \times B}{1 + (\mu B)^2}\right), \tag{3.29}$$

If E and B are in phase ($E \times B = 0$), then the velocity equation is simplified into:

$$v = \mu E\left(\frac{1}{1 + (\mu B)^2}\right), \tag{3.30}$$

which is obviously smaller than v without the presence of magnetic field. Smaller v means charges move slower and hence result in a smaller current. Obviously the charge velocity (and hence current) increases, if it is at 90 ° to the magnetic field.

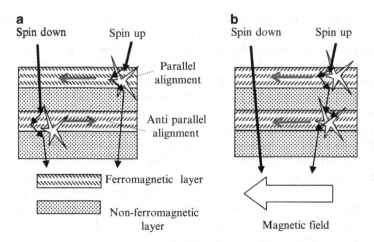

Fig. 3.17 GMR effect (**a**) before and (**b**) after applying the magnetic field

Hence, magnetoresistors depend on the direction of magnetic field. Considering this effect, it is possible to develop sensors for monitoring the orientation of magnetic fields based on the magnetoresistance. In these sensors, the resistance changes, when the resistance of a magnetoresistor alters upon the change of a magnetic field polarity.

Magnetoresistance for natural materials are not high and does not exceed values required for detecting very small magnetic pixels in high density disk drives. The effect has become much more prominent after the discoveries of *anisotropic magnetoresistance* (*AMR*) and *giant magnetoresistance* (*GMR*).

AMR is an effect that is only found in ferromagnetic materials (materials with larger μ), where the electrical resistance increases when the direction of current is parallel to the applied magnetic field. The change in the material's electrical resistivity depends on the angle between the directions of the current and the ferromagnetic material's magnetization.

The change of resistance in GMR is much larger than AMRs so GMRs play a significant role in today's sensing. They are largely used in read heads of the magnetic disks for computers information storages for sensing magnetic fields, among other sensing applications. The progress in the development of GMR sensors has allowed the fabrication of magnetic sensors for increasingly smaller magnetic beads in recent years, which has resulted in the possibility of increasing hard disk capacities. GMR was first discovered independently in 1988 by research teams led by Peter Grünberg of the Jülich Research Centre [3] and Albert Fert [4] of the University of Paris-Sud. It relies on the quantum nature of materials and is observed in layered structures that are composed of alternating ferromagnetic and nonmagnetic metal layers, with the thickness of the nonmagnetic layer in the nanometer order.

The effect is illustrated in Fig. 3.17. Electron scattering at the ferromagnetic/nonmagnetic interfaces depends on whether spin of electron passing through the material is parallel or anti-parallel to the magnetic moment of the layer.

Before applying the electric field the ferromagnetism of the layers are in parallel and anti-parallel intervals.

The electron has two spin values (up and down). The first magnetic layer allows electrons in only one spin state to pass through easily. Without any magnetic field, the magnetic moments of the adjacent layers are not aligned, then electrons with either spin cannot get through the structure easily and the electrical resistance is high. When a magnetic field is applied, the magnetic moments of the adjacent layers are aligned, then electrons with that spin value can easily pass through the structure, and the resistance is low for those electrons.

3.18 Doppler Effect

The *Doppler effect* concerns the apparent change in wave's frequency as a result of the observer and the wave source moving relative to each other. If the observer and wave source are moving toward each other, the wave appears to increase in frequency and is said to be *hypsochromically* (or blue) shifted (Fig. 3.18). Conversely, if the wave source and observer are moving away from each other, then the wave appears to decrease in frequency and becomes *bathochromically* (or red) shifted.

The observed Doppler shift in frequency is given by:

$$f_{\text{observed}} = \left(\frac{v}{v + v_{\text{source}}} \right) f_{\text{source}}, \tag{3.31}$$

where v is the speed of the wave in the medium, v_{source} is the speed of the source with respect to the medium, and f_{source} is the frequency of the source wave. If the wave source approaches the observer, then v_{source} is negative, and conversely, if the wave is receding, then it takes on a positive value. A familiar example of the Doppler effect include the changing pitch of an ambulance siren as it approaches and then drives past the observer.

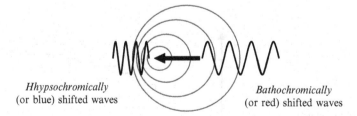

Hhypsochromically
(or blue) shifted waves

Bathochromically
(or red) shifted waves

Fig. 3.18 Hypsochromic and bathochromic frequency shifts occurring as a result of the Doppler effect

Examples of the Doppler effect in sensing include speed monitoring devices and ultrasounds. The Doppler effect also plays an important role in radar and sonar detection systems. Additionally, hypsochromic and bathochromic shifts are commonly used in velocity measurement of large and distant bodies such as stars, galaxies, and gas clouds as their motion and spectrum can be studied with respect to the observer.

3.19 Barkhausen Effect

In 1919 Heinrich Barkhausen found that applying a slowly increasing, continuous magnetic field to a ferromagnetic material causes it to become magnetized, not continuously, but in small steps (Fig. 3.19). These sudden and discontinuous changes in magnetization are a result of discrete changes in both the size and orientation of ferromagnetic domains (or of microscopic clusters of aligned atomic magnets) that occur during a continuous process of magnetization or demagnetization. This effect generally should be reduced in the operation of magnetic sensors, as it appears as a step noise in measurements.

3.20 Nernst/Ettingshausen Effect

Walther Hermann Nernst and Albert von Ettingshausen discovered that an electric field is produced across a conductor or semiconductor, when it is subjected to both a temperature gradient and a magnetic field. The direction of this electric field is mutually perpendicular to both the magnetic field and the temperature gradient. The effect can be quantified by the Nernst coefficient $|N|$ as:

$$|N| = \frac{E_y/B_z}{dT/dx}.$$ (3.32)

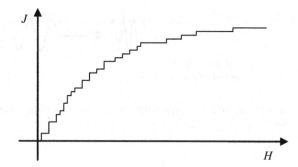

Fig. 3.19 Exaggerated demonstration of Barkhausen jumps in magnetization (J) curve as a function of magnetic field intensity (H) in a ferromagnetic material

This equation described that, if the magnetic field component is in the z-direction, B_z, then the resulting electric field component will be in the y-direction, E_y, when subjected to a temperature gradient of dT/dx.

Due to the thermoelectric effect, electron and holes move along temperature gradients. When a magnetic field, transversal to this temperature gradient, is applied, the electron and holes experience a force perpendicular to their direction of motion, and therefore, a perpendicular electric field is generated. The Nernst effect can be seen in many semiconductors. In spite of its potential, this effect is yet to be fully investigated for its application in sensing.

3.21 Faraday and Voigt Rotation Effects

Discovered by Michael Faraday in 1845, it is a *magneto-optic effect* in which the polarization plane of an electromagnetic wave propagating through a material becomes rotated when subjected to a magnetic field that is parallel to the propagation direction. This rotation of the polarization plane is proportional to the intensity of the applied magnetic field. Faraday rotation effect is due to the alteration of the dielectric tensor in a material as a result of magnetization.

The effect was the first experimental evidence that light is an electromagnetic wave and was one of the foundations on which James Clerk Maxwell developed his theories on electromagnetism. The angle of rotation is defined by equation (Fig. 3.20):

$$\theta = VBl, \tag{3.33}$$

where B is the magnetic flux density, V is the *Verdet constant*, and l is the length of the material through which the light is passing.

The Faraday effect has been widely used in high frequency measurement and sensing systems. It is commonly implemented for the remote sensing of magnetic

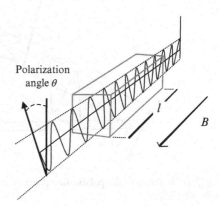

Fig. 3.20 Rotation of the polarization plane as a result of the Faraday effect

fields. It is increasingly used for measuring the polarization of electron spins. Additionally, Faraday rotators are regularly employed in telecommunication systems for amplitude modulation of optical waves and in optical circulators.

The Verdet constant is a figure of merit used that shows the strength of the Faraday effect in different materials. A common magneto-optical material for field sensing is *terbium gallium garnet* (*TGG*), which has a Verdet constant as large as -40 rad T^{-1} m^{-1}. Incorporating such materials, Faraday rotation angles of over 45 ° can be achieved, which allows the construction of Faraday rotators, devices which transmit light in only one direction.

Voigt effect is similar to Faraday rotation effect. However, Faraday effect is linearly proportional to the applied magnetic field, while Voigt effect shows a quadratic dependence. This effect was discovered in 1902 and is observed in many materials in both solid and vapor forms.

3.22 Magneto-Optic Kerr Effect

In 1877, John Kerr discovered that the polarization plane of a light beam incident on a magnetized surface is rotated by a small amount after it is reflected from that surface. This is because the incoming electric field, E, of the light exerts a force, F, on the electrons in the material, and consequently they vibrate in the plane of polarization of the incoming wave. If the material has some magnetization, M, then the reflected wave will gain a small electric field component (called the Kerr component, K), which is a vector added to the reflected electric field wave vector. As a result, the reflected wave is rotated with respect to the incident wave, as demonstrated

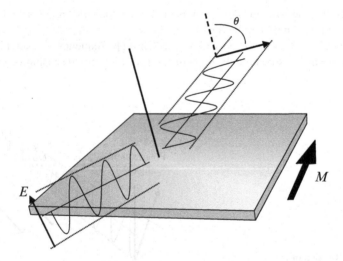

Fig. 3.21 Rotation of the polarization plane on a magnetized surface as a result of the magneto-optic Kerr effect

in Fig. 3.21. The amount of rotation depends on the magnitude of M. Similar to Faraday rotation effect, the magneto-optic Kerr effect (MOKE) also occurs when the magnetization in the material produces a change in its dielectric tensor.

The Kerr effect can be used to fabricate sensors for various applications. A Kerr Microscope takes advantage of the MOKE in order to image differences in magnetic orientation and hence mapping them across ferroelectric materials. It is possible to develop pressure sensor based on MOKE. The pressure difference across the diaphragm, which is covered with a magnetic material, causes deflection and thus stress in the layer. This leads to a change in the magnetic properties of the thin-film, which can be measured using MOKE. Additionally, MOKE can also be used for reading information stored in magnetic disks.

3.23 Kerr Effect

Discovered by John Kerr in 1875, it is an electro-optic effect. Upon exposure to an electric field some materials become birefringent and show different refractive indices for light polarized parallel to and perpendicular to this applied field. Here, birefringence is induced electrically in isotropic materials. When an electric field is applied to liquid or gases, their molecules (which have electric dipoles) may become partly oriented with the field. This renders the substance anisotropic and causes birefringence in the light traveling through it. However, only light passing through the medium normal to the electric field lines experience this birefringence, and it is proportional to the square of the electric field. Hence the effect is different for various polarizations. The amount of birefringence due to the Kerr effect is proportional to the square of the applied electric field, E, given by:

$$\Delta n = n_{\mathrm{o}} - n_{\mathrm{e}} = \lambda_{\mathrm{o}} K E^2,\tag{3.34}$$

where K is the Kerr constant and λ_{o} is the wavelength of light. The two principal indices of refraction, n_{o} and n_{e}, are the ordinary and extraordinary indices, respectively.

The effect is utilized in many optical devices such as shutters, monochromators, and modulators. Polar liquids such as nitrotoluene ($C_7H_7NO_2$) exhibit very large Kerr constants, which are frequently used in light modulators at frequencies exceeding 10 GHz. However, the drawback is the relative weakness of Kerr effect which requires voltages as high as 30 kV for complete modulation.

3.24   Pockels Effect

Pockels effect is a similar effect to the Kerr effect, with the difference being the birefringence is directly proportional to the electric field, not its square (as in the Kerr effect). This effect is seen in some of solid crystals that lack a center of symmetry (20 out of the 32 classes). In contrast to Kerr effect, Pockels effect can produce the refractive index changes at much lower voltages.

Pockels effect is seen in many crystals such as *lithium niobate* ($LiNbO_3$) and *gallium arsenide* (*GaAs*) and has variety of applications in optics and sensors. Incorporated with a polarizer, a Pockels cell can be used as an optical switch alternating between no rotation and 90 ° rotation in nanoseconds. Pockels cells are also regularly used in electro-optic probes.

3.25   Summary

Some of the physical transduction effects that are commonly used in sensing systems were presented in this chapter. This includes electromagnetic, dielectric, permeability, photoelectric, photoconductive, photovoltaic, photodielectric, photoluminescence, electroluminescence, Hall Effect, thermoelectric, thermoresistive, piezoresistive, piezoelectric, pyroelectric, magnetostriction, mangnetoresistance, Doppler, Barkhausen, Nernst/Ettingshausen, Faraday and Voigt rotation, magneto-optic Kerr, Kerr, and Pockels effects.

In the next chapter, it will be shown how some of these effects are incorporated into sensing platforms.

References

1. Peng XJ, Du JJ, Fan JL, Wang JY, Wu YK, Zhao JZ, Sun SG, Xu T (2007) Selective fluorescent sensor for imaging Cd^{2+} in living cells. J Am Chem Soc 129:1500–1501
2. Clark AE, Belson HS (1972) Giant room temperature magnetostriction in $TbFe_2$ and $DyFe_2$. Phys Rev B 5:3642–3644
3. Grünberg P, Schreiber R, Pang Y, Brodsky MB, Sowers H (1986) Layered magnetic-structures – evidence for antiferromagnetic coupling of Fe layersacross Cr interlayers. Phys Rev Lett 57:2442–2445
4. Baibich MN, Broto JM, Fert A, Nguyen Van Dau F, Petroff F, Eitenne P, Creuzet G, Friederich A, Chazelas J (1988) Giant magnetoresistance of (001) Fe/(001) Cr magnetic superlattices. Phys Rev Lett 61:2472–2475

Chapter 4
Transduction Platforms

Abstract The major transduction platforms utilized in sensing applications will be presented and discussed. The focus is on systems, which can be fabricated and set up utilizing available industrial development processes.

4.1 Introduction

In the previous chapter, the major physical transduction effects were presented. In continuation, this chapter will focus as how those effects can be incorporated and used in transduction platforms that are used for sensing. The platforms will be described one by one and examples will be provided as how they are used for sensing.

4.2 Conductometric and Capacitive Transducers

Conductometric (or *resistive*) and *capacitive* transducers are among the most commonly utilized devices in sensing applications. This is largely due to their inexpensive fabrication and simple operation. They involve placing a material between conducting electrodes. When these sensors are exposed to stimuli, the electrical conductivity or capacitance is measured to generate output signals associated with the target stimuli. A typical setup with only two electrodes and a sensitive layer is shown in Fig. 4.1a. In order to increase the effect of sensitive layer, the area between the electrodes can be increased by using inter-digital configurations (Fig. 4.1b).

K. Kalantar-zadeh, *Sensors: An Introductory Course*,   63
DOI 10.1007/978-1-4614-5052-8_4, © Springer Science+Business Media New York 2013

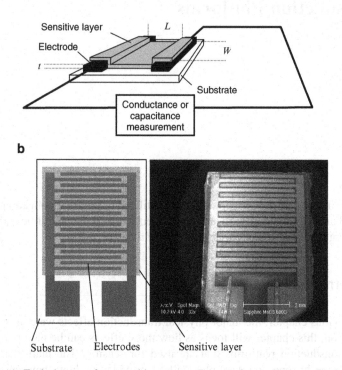

Fig. 4.1 (**a**) Typical setup for capacitive or conductometric transducers. (**b**) Inter-digital transducers (IDTs). *Bottom left*: schematic; *Bottom right*: electron microscopy image

4.2.1 Conductometric Sensors

In conductometric measurements, conductance of a sensitive material, which is placed between electrodes, is measured. Either a voltage is applied across the electrodes, which generates a current or a current is forced through the device to generate a voltage. The resultant current or voltage depends on the conductivity of the sensitive material, which alters as a result of its exposure to the target measurand.

When a current passes through the device, the current density, J, electric field, E, and electrical conductivity, σ, are related through Ohm's law:

$$J = \sigma E. \tag{4.1}$$

The relation between voltage and electric field in a two electrode device, with parallel electrodes of distance L, is given as follows:

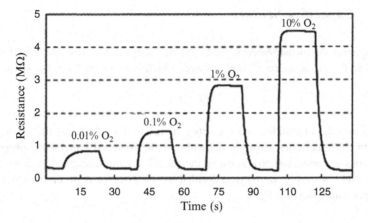

Fig. 4.2 Dynamic response of the SnO_2 thin film conductive transducers to O_2 gas in nitrogen gas ambient (reproduced from [1] with permission)

$$E = \frac{V}{L}, \tag{4.2}$$

which assists in transforming it into a more common form of Ohm's law is as:

$$V = IR, \tag{4.3}$$

where V is the voltage, I is the current ($J = I/Wt$), and R is the electrical resistance. As a result, the relation between the resistance and conductivity is:

$$R = \frac{1}{\sigma} \frac{L}{tW}, \tag{4.4}$$

in which W is the width of the sensitive layer and t is the thickness of the layer. Sheet resistance is defined as $R_s = 1/t\sigma$.

Many materials exhibit nonlinear electrical conductivity, and therefore careful consideration of the biasing conditions must be made during a measurement. Generally the applied voltages and currents are selected such that the conductivity remains in a relatively linear region of the calibration curve.

Conductometric devices are generally the most facile sensors to fabricate with widespread applications in various industries. A large number of different types of sensors such as thermistors, photoconductors, conductometric biosensors, and semiconductor gas sensors are based on these simple electrodes. As an example, Fig. 4.2 shows the change in electrical resistance of a tin oxide (SnO_2) thin film when exposed to different concentration of oxygen gas (O_2) in nitrogen gas [1]. As can be seen, the electrical resistance alters in the presence and absence of O_2 gas of different concentrations from the baseline of ~0.25 $M\Omega$ to the values 0.78, 1.5, 2.8, and 4.5 $M\Omega$.

4.2.2 Capacitive Sensors

For capacitance measurements, build up of charge, Q, across the electrodes is related to the capacitance, C, and voltage, which is given by the relationship:

$$Q = CV. \tag{4.5}$$

Between the electrodes is a dielectric material. The electrical field that arises between the electrodes strongly depends on the material's dielectric properties. Analogous to conductometric sensors, in a parallel electrode capacitor, in which the electric field is simply the voltage divided by the distance between the electrodes, the capacitance is given by:

$$C = \varepsilon \frac{Wt}{L}, \tag{4.6}$$

where Wt is the electrode area, L is the distance between the electrodes, and ε is the dielectric constant.

Capacitive devices are extensively used as fluid gauge, humidity, pressure, and non-contact sensors. The schematic of fluid gauge operation is shown in Fig. 4.3a. By placing two metal electrodes in liquid, a capacitor is formed whose capacitance is proportional to the permittivity of fluid between. Depending on how deep the electrodes have been emerged (the fluid level) the capacitance changes. The increase in capacitance is associated with the increase of the fluid level. Obviously for correct measurements, the conductivity of the fluid should also be considered. The other common types of capacitive devices are the ones, which are employed for pressure sensing. One configuration of capacitive pressure sensors is demonstrated in Fig. 4.3b. It is made of two metallic electrodes, forming a capacitor, and a diaphragm, which is attached to one of the electrodes. The metallic electrode, which is attached to the diaphragm, is flexible to move together with it. The other side of diaphragm is in contact with the fluid (or gas) media whose pressure should be measured. Change of pressure bends the diaphragm that consequently changes the capacitance. Another category of capacitive sensors are non-contact type sensors in which the surface of electrodes is covered with an insulating material as schematically shown in Fig. 4.3c. When an external object in the vicinity of the electrodes, the permittivity of the media and hence, the capacitance change. Non-contact capacitive sensors are extensively used in fingerprint imagers. Such systems are made of an array of capacitive sensors, consist of adjacent plate capacitors, with resolutions as high as 500 dots per inch or more.

Both conductometric and capacitive measurements can be carried out under either DC or AC conditions. A material's conductivity and dielectric properties generally exhibit strong frequency dependence and therefore operating frequency can have a major impact on measurements.

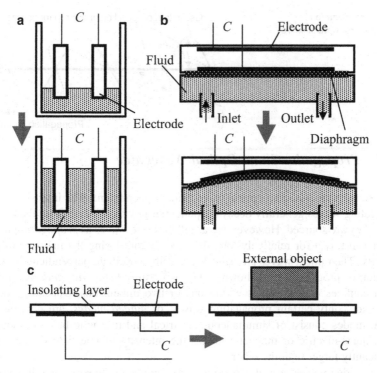

Fig. 4.3 Different types of capacitive sensors: (**a**) fluid gauge, (**b**) pressure, and (**c**) non-contact

4.3   Optical Waveguide Based Transducers

Optical waveguide based transducers are among the most utilized transducers in sensing. Such sensors utilize interactions of optical waves, generally in the visible, infrared, and ultraviolet regions, with the measurand. These interactions cause the properties of the waves such as intensity, phase, frequency, and polarization to change and these changes are then associated with the target measurand.

In this section, some typical optical waveguide based transducers, which are commonly used in sensing, are presented. These are categorized into two major types: transducers based on propagation of optical waves *confined into waveguides* and transducers based on *surface Plasmon waves*. Prior to presenting them, information regarding light propagation and the sensitivity of such waveguides is provided.

The readers should consider that there are many types of optical sensors, which are not based on waveguiding structures; rather they make use of *spectroscopy* for characterizing target analytes. Such systems will be discussed in the next section.

Fig. 4.4 Propagation
of a transverse optical wave

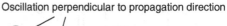

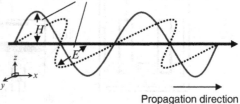

4.3.1 *Propagation in an Optical Waveguide*

Many sensors are based on optical wave guiding phenomena. The intensity of such propagating waves generally decays exponentially with distance from the point at which they are sourced. However, we usually refer to waves as *propagating waves*, if they can travel for relatively long distances before losing the majority of their intensity. The optical waves are *transverse*, as they oscillate perpendicularly to the direction of propagation, as shown in Fig. 4.4. An *optical waveguide* is a media, which confines optical waves within one or two dimensions. Depending on the waveguide, only certain propagating waves, or *guided modes*, are possible. All optical modes consist of simultaneous electrical and magnetic field components, but in quasi-electric or magnetic modes the intensity of one field component is significantly larger than the other.

The guided waves in a *planar optical waveguide*, confined in two dimensions, are either TM (*transverse magnetic* or *p-polarized*) or TE (*transverse electric* or *s-polarized*). At any time, *t*, a normalized single frequency (*monochromatic*) propagating wave in the *x* direction can be represented by the function:

$$f(x,t) = e^{i(k_x x - \omega t)}, \tag{4.7}$$

where $\omega = 2\pi f$ is the angular frequency and k_x is the propagation constant (also called *wavenumber*) in *x* direction. Similarly the propagation in the *y* and *z* directions can be described by the wavenumbers k_y and k_z in those directions, respectively.

The schematic of a simple two-dimensional, planar, waveguide is shown in Fig. 4.5a. The cross-sectional depiction of this planar optical waveguide is also presented in Fig. 4.5b. The waveguide is made of a layer deposited on a substrate. When light is coupled into the in between layer, it can only stay confined in this layer, if the refractive index of this layer (N_l) is larger than its top cladding media (N_c) and bottom substrate (N_s). In sensing applications, the top media is generally where the target material is placed (Fig. 4.5b—the external object). The parameters of the sensors should be designed in a way that the propagating wave mode distribution deforms to produce penetrating tail into the target sensing media. The change of penetration tail (evanescence tail) changes in the intensity, velocity, and/or phase of the propagating wave. The change in the output signal profile at the other end of the waveguide corresponds to the presence of a target measurand.

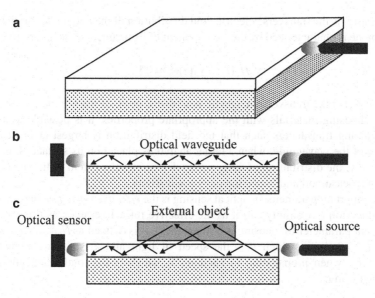

Fig. 4.5 (**a**) Schematic of a two-dimensional waveguide. (**b**) Cross-sectional view. (**c**) Cross-sectional view when an external object is in the vicinity of the waveguide

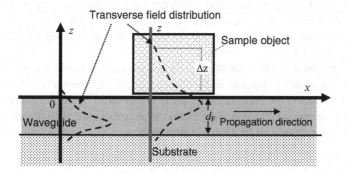

Fig. 4.6 Distribution of an evanescent wave in z direction. Δz is the penetration of the field into the sample object

The wave propagates in the x direction and, for simplicity, the waveguide can be considered to have either infinite or finite dimensions in the y direction. The thickness of the waveguide, d_f, is finite (Fig. 4.6). Also seen in this figure is the field distribution of the mode component in z direction (dotted lines). Engineering this distribution allows the device to be optimized for sensing applications. The TE_m mode (m is the number of mode) is characterized by the y component of its electric field and is given by:

$$E_y(t) = u_m(z)e^{i(k_x x - \omega t)}, \tag{4.8}$$

where $u_m(z)$ is the transverse electric field distribution of the mth mode. Similarly, a TM$_n$ mode is characterized by the y component of its magnetic field and is given by:

$$H_y(t) = v_n(z)e^{i(k_x x - \omega t)}, \tag{4.9}$$

where $v_n(z)$ is the transverse magnetic field distribution of the nth mode.

By choosing materials with the appropriate properties, it is possible to design waveguiding transducers such that the field distribution is largest at or near the surface of the waveguide, when a target analyte is placed on its surface (Fig. 4.6). Ultimately, the distribution of these waves determines the extent of the interaction with the measurand and hence the device sensitivity.

An important parameter in optical sensing is the *effective refractive index*, N_{eff} in the x direction (we simply call it as the effective refractive index as it is the most important parameter for sensing in waveguides). It is defined as $N_{\text{eff}} = k/k_x$, where $k = \omega/c = 2\pi/\lambda$, in which c is the speed of light and λ is the wavelength of the waves, when propagating in vacuum. k, which is the total wavenumber, is obtained using:

$$|k| = \sqrt{|k_x|^2 + |k_z|^2}, \tag{4.10}$$

k is a vector that describes the rate at which the wave propagates in its direction. The effective refractive index, and hence k, is a function of the polarization of the propagating wave, mode number, the waveguide thickness d_f, and the refractive indices of the media.

As was briefly described, the propagating waves can be *evanescent* in z direction. This means that their z component distribution decays exponentially, when entering the sample media or the substrate. The decay is described by the *penetration depth*, Δz, into the sample media (Fig. 4.6). From the penetration depth, the field distribution in the z direction can be approximated with:

$$u_m(z) = u_m(0)e^{\frac{-z}{\Delta z}}, \tag{4.11}$$

$$v_m(z) = v_m(0)e^{\frac{-z}{\Delta z}}. \tag{4.12}$$

Here $1/\Delta z$ is equal to the real part of the wavenumber, k_z, in the z direction in the sample media. By multiplying Δz by k the effective refractive index in the z direction is obtained as $2\pi\Delta z/\lambda$. Considering that the propagation wavenumber in the x direction, k_x, should be equal for all media (wave front propagates with one velocity in the x direction) the magnitude of Δz can be calculated as [2]:

$$|\Delta z| = \frac{\lambda}{2\pi}\left(N_{\text{eff}}^2 - N_C^2\right)^{-\frac{1}{2}}, \tag{4.13}$$

where N_C is the refractive index of sample media. The effective refractive index in the sample media is:

$$\left(\frac{2\pi}{\lambda}\right) \times N_{\text{eff}} = k, \qquad (4.14)$$

If the refractive index of the sample media changes, then the penetration depth will also change according to (4.13). This in turn results in a measurable change in the field distribution and is the basis of affinity sensing with optical waveguides.

When fabricating sensors based on optical waveguide transducers, materials should be chosen based on how the penetration depth of the system can be manipulated and affected in the presence of a target object. The control of the penetration depth length is important as it can be used for adapting the system for the optimal detection of analyte molecules, whose dimensions are of various orders. For instance, analyte materials, such as proteins and DNA strands, can be detected using optical waveguide platforms by creating penetration lengths in the orders of tens to hundreds of nanometers.

4.3.2 Sensitivity of Optical Waveguides

The sensitivity of an optical waveguide based sensor strongly depends on the interaction between the measurand and the surface confined guided mode in the sample media. Analyte molecules may diffuse into or out of the evanescent region, they may become immobilize onto the boundary, or they may move along the surface by forces such as convection and flow. Each of these interactions can change the effective refractive index, and as a result, produce an optical response. The change in effective refractive index can be calculated using the perturbation theory. For TM modes, the result is expressed as [2]:

$$\Delta(N_{\text{eff}}^2) = \frac{\frac{\int_{-\infty}^{+\infty} \Delta\varepsilon(z) \left(\frac{dv(z)/dz}{\varepsilon(z)}\right)^2 dz}{k^2} - N_{\text{eff}}^2 \int_{-\infty}^{+\infty} \Delta\left(\frac{1}{\varepsilon(z)}\right) |v(z)|^2 dz}{\int_{-\infty}^{+\infty} \frac{|v(z)|^2}{\varepsilon(z)} dz}, \qquad (4.15)$$

and for TE modes it is:

$$\Delta(N_{\text{eff}}^2) = \frac{\int_{-\infty}^{+\infty} \Delta\varepsilon(z) |u(z)|^2 dz}{\int_{-\infty}^{+\infty} |u(z)|^2 dz}, \qquad (4.16)$$

which can be calculated using various computational methods. $v(z)$ and $u(z)$ are field distributions for the TM and TE modes, respectively.

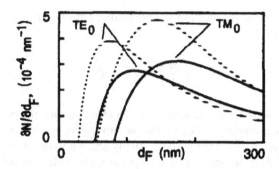

Fig. 4.7 Sensitivity (the change of refractive index with reference to the thickness of the protein layer) of the TE_0 and TM_0 modes (reproduced from [2] with permission)

From (4.15) and (4.16) the sensitivities for both optical waveguide and *surface Plasmon* (*SP*) based sensors can be obtained by measuring the change in effective refractive index with respect to the change of the waveguide thickness ($\partial N_{eff}/\partial d_F$) and with respect to the change of refractive index in the sample media ($\partial N_{eff}/\partial N_C$). These two are important sensing parameters: the former describes how much material is attached to the surface of the waveguide (e.g., a layer of a biomaterial attached onto the optical waveguide surface and sensed) and the latter describes how the refractive index of the target media is changed (e.g., the attached target is a biomaterial with a refractive index, which is different from the media). It is a common approach that for developing biochemical and chemical sensors from optical waveguide platforms, a layer that is chemically sensitive to the target measurand is deposited on top of the waveguide.

An example of an optical waveguiding biosensor's sensitivity is shown in Fig. 4.7. The sensitivity of TE_0 (zeroth order transverse electric mode—also referred to as the TE fundamental mode) and TM_0 (zeroth order transverse magnetic mode—also referred to as the TM fundamental mode) modes is calculated for the condition in which a layer of protein that has a refractive index of approximately 1.45 is added on top of the guiding layer [2]. In this example, the refractive indices of substrate and the guiding layer are 1.8 and 1.47 for the solid lines and 2.0 and 1.46 for the dashed lines, respectively. As can be seen, the larger the difference between the refractive indices of the substrate and guiding layer, the larger the sensitivities will be. The refractive index of the sample media is 1.33.

Generally, the sensitivity can be increased by reducing the waveguide's thickness, as it allows larger evanescence tail. Producing a single propagation mode is another option. Having single mode of operation, energy distribution is limited within a selected mode and any perturbation affects the single mode rather than causing exchange of energy between modes. The smaller thickness waveguide and single mode conditions more likely guarantee the confinement of the energy at near the surface of the waveguide.

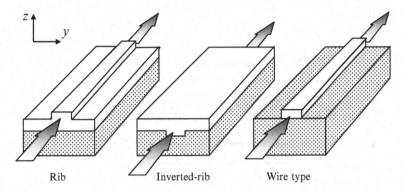

Fig. 4.8 Different configurations of optical waveguides

4.3.3 Optical Fiber Based Transducers

Optical waves can be confined into quasi-one-dimension: structures such as optical fibers rather than planar arrangements. Some of the most common of such configurations are *rib*, *inverted-rib* , and *wire type waveguides* (Fig. 4.8). In these waveguides, in addition to confinement in the z direction, the optical waves are also confined in y direction. Light waves propagate along the rib, inverted-rib, and wire structure, provided that the refractive index of the core is greater than that of the claddings (surroundings). Obviously, by placing the target measurand in the vicinity of the optical waveguide, interaction with this medium alters the optical waves' properties such as phase, velocity, and amplitude, when compared to the light originally propagating. The more the wave penetrates into this medium the more this sensing effect can be observed. For chemical and biochemical sensing, the exposed core may also be covered with a sensitive layer to further provide selectivity to a target analyte.

Optical fibers are another type of such waveguides, in which light is confined within two dimensions. They consist of a core which is covered by a cladding. Optical fiber based transducers have several configurations: two of the most popular are shown in Fig. 4.9. In the first, part of the fiber's cladding is removed, exposing the core directly to the sample medium. In the second configuration, the end of the fiber is exposed to sample medium, and the optical waves reflecting or passing through are analyzed. Once again, selectivity to a target can be obtained by depositing a sensitive layer at the fiber's end.

4.3.4 Interferometric Optical Transducers

Interferometric optical transducers are based on the constructive and destructive interference of optical waves. In general, waves traveling through a waveguide are

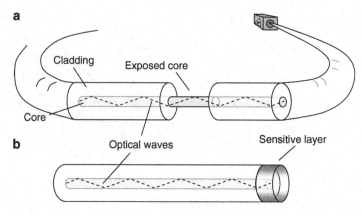

Fig. 4.9 Schematic of optical fibers as sensors (**a**) removed cladding (**b**) reflective or passing through type

split into two or more beams of equal intensity. After splitting, one of the beams travels unperturbed through the waveguide, whilst the other travels through a waveguide that may be exposed to the sample media. Depending on how the second beam is affected by the media, the light beams are recombined either destructively or constructively, and the optical properties of this recombined beam (e.g., intensity, wavelength, phase, etc.) are reordered.

Interferometric optical sensors are ideal for realizing on-chip optical sensors, as they can be fabricated using conventional microfabrication techniques. They have standard input and output for beam coupling, capability for differential on-chip referencing, and are highly stable. However, their major drawbacks are their rather large dimensions and fabrication costs.

The most common interferometric sensors are based on Mach–Zehnder configurations. Schematic of a typical *Mach–Zehnder interferometer*, used for sensing, is shown in Fig. 4.10.

In a Mach–Zehnder interferometer, the light source is coupled at the input. The light wave at the Y junction splits into two parts. If it is a symmetric interferometer, then:

$$F_{\text{total}}(x,y) = \frac{1}{2}F\left(x + \frac{1}{2}S, y\right) + \frac{1}{2}F\left(x - \frac{1}{2}S, y\right), \qquad (4.17)$$

where S is the distance between the two arms, and $F(x, y)$ is the function that represents the intensity of light at (x, y). A section of cladding in one of the waveguide arms can be removed, and the wave may be directly exposed to the sample media. Also to increase the sensitivity, a sensitive layer may be placed, where the cladding was removed. After the beam interacts with the analyte, a phase shift between the light waves traveling through both arms occurs. If the propagation light in each of the arms is single mode, the output field is given by:

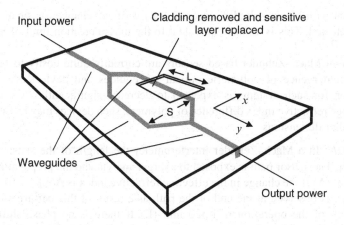

Fig. 4.10 Schematic of a Mach–Zehnder sensor

$$F_{total}(x, y) = \frac{1}{2}F(x + \frac{1}{2}S, y)e^{i\Delta\varphi} + \frac{1}{2}F(x - \frac{1}{2}S, y), \qquad (4.18)$$

in which the phase difference between the two arms is calculated as:

$$\Delta\varphi = \left(\frac{2\pi}{\lambda}\right)\Delta N_{eff} L, \qquad (4.19)$$

where ΔN_{eff} is change in the effective refractive index, λ is the wavelength of light, and L is the length of the sensitive layer. Eventually both light beams are recombined in the second Y junction. As a first-order approximation, we assume that the amplitudes of the two beams are still equal. In this case, the two beams are out of phase by $\Delta\varphi$, and the intensity of the resulting beam is related to the phase difference as:

$$I = \frac{1 + \cos(\Delta\varphi)}{2}. \qquad (4.20)$$

Obviously the output signal of such a system can vary between zero, when the two signals are 180° out of phase, and its maximum, when they are in phase. The beam intensity change ΔI accounts for the sensitivity, and the sensitivity can be defined as:

$$\frac{\partial I}{\partial d_F} + \frac{\Delta I}{\partial N_C} = \frac{\partial I}{\partial \varphi}\frac{\partial \varphi}{\Delta N_{eff}}\left(\frac{\Delta N_{eff}}{\partial d_F} + \frac{\Delta N_{eff}}{\partial N_C}\right), \qquad (4.21)$$

where d_F and N_C are the changes in the thickness and the upper cladding refractive index of the waveguide, where the target sample is placed, respectively. It has been shown that with Mach–Zehnder sensors refractive index changes as small as 10^{-6} can be detected. This is almost equivalent to the mass detection limit of less than 1 pg mm^{-2}.

Although Mach–Zehnder based sensors are currently quite costly to fabricate, with the emergence of polymeric optical waveguides and devices, the cost of fabrication for such systems is expected to decrease significantly. Additionally, due to large refractive index difference of polymers, such devices may be fabricated with smaller dimensions.

Question 1: In a Mach–Zehnder interferometer the length of the sensor path is $L = 1$ cm. The sensor path is exposed to a target analyte and the propagating wave is affected. (A) If the change in the effective refractive index is $\Delta N_{eff} = 10^{-5}$, what phase shift is observed at the end of the path as a result of this perturbation? The wavelength of the operation is 1,520 nm. (B) If there is no phase shift in the reference path and there is no attenuation in paths, what is the ratio of the output power-to-input power?

Answer:

(A) In the sensing path of a Mach–Zehnder interferometer, the change in the phase shift is calculated using:

$$\Delta\varphi = \left(\frac{2\pi}{\lambda}\right)\Delta N_{eff}L = \left(\frac{2\pi}{1,520 \times 10^{-9}\,(m)}\right) \times 10^{-5} \times (10^{-2}\,(m))$$
$$= 0.413\,(rad) = 23.68°.$$

(B) The input signal has the intensity magnitude of $I_{in} = |I|$, which is equally split between the reference and sensor paths. In the sensor path, if the input is $|I/2|$, the output will be $|I/2| \times \cos(\Delta\varphi)$. As a result, the total signal at the output will be $|I/2| \times (1 + \cos(\Delta\varphi))$. As $\cos(23.68°) = 0.916$, the output power is $I_{out} = 0.958|I|$. As a result, $I_{out}/I_{in} = 0.958$.

4.3.5   Surface Plasmon Resonance Transducers

If waveguide thickness, d_f, decreases to become a thin sheet of infinitely small thickness, then the waveguide gradually changes to boundary-like thin film between two media: the substrate and the sample (Fig. 4.11). Also, if this waveguide is metallic, then the TM mode waves may become trapped near and around this surface and propagate along the boundary. If this occurs, then these waves are known as surface Plasmon waves, and their excitation by light is called *surface Plasmon resonance* (*SPR*). Since the SPR wave can exist on the boundary of the metal and the target medium, its distribution in the z direction is very sensitive to any change of this boundary. As a result, the SPR systems can be efficiently used as affinity sensors that are incorporated for the recognition of target molecules adsorption to the metal surface.

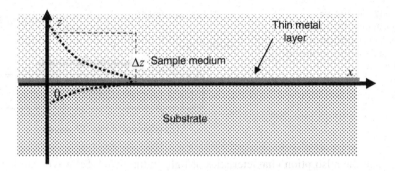

Fig. 4.11 A SP waveguide consisting of a thin metal film between the sample medium and substrate

For the SPR to occur a thin metal surface is needed for which the real part of its relative permittivity is negative:

$$\varepsilon_M = \varepsilon'_M + i\varepsilon''_M \text{ with } \varepsilon'_M < 0. \tag{4.22}$$

The metallic layer has a k_M wavenumber associated with ε_M, the refractive index of the metal. Refractive index is equal to the square root of relative permittivity ($n^2 = \varepsilon$). Ignoring the imaginary value of (4.22), the majority of the metal effective refractive is influenced by ε'_M, which is its real component. The SP wave, which is a TM mode wave, is defined by general equations (4.9) and (4.12) for waveguides. As a result, the effective refractive index obtained as [2]:

$$N_{\text{eff}} = \left(\varepsilon'_M{}^{-1} + N_C{}^{-2}\right)^{-\frac{1}{2}}. \tag{4.23}$$

For a SP wave two conditions are required: (1) ε'_M has to be negative and (2) the magnitude of the real part of the metal's permittivity must be greater than the square of the refractive index of the sample medium:

$$|\varepsilon'_M| > N_C^2. \tag{4.24}$$

Having these two conditions, (4.23) results in an imaginary effective index, hence an imaginary propagation wavenumber, which is associated with a lossy wave in the x direction.

The SP wave's field distribution in the sample media is evanescent in the z direction. The penetration depth can be obtained by combining (4.13) and (4.23) (rearranging (4.23) to $N_{\text{eff}}{}^{-2} - N_C{}^{-2} = \varepsilon'_M{}^{-1}$) which results in:

$$\Delta z = \left(\frac{\lambda}{2\pi N_C N_{\text{eff}}}\right)(-\varepsilon'_M)^{\frac{1}{2}}. \tag{4.25}$$

Fig. 4.12 Schematic demonstrating SP waves coupling via a planar optical waveguide—cross section of the system

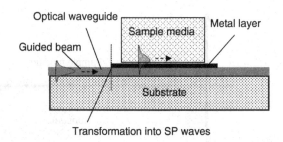

Transformation into SP waves

Due to the adsorption characteristics of light in the metal, SP waves are highly attenuated in the visible spectrum. The attenuation is attributed to the relatively large imaginary part of the metal's permittivity. The intensity decays exponentially as $e^{-\alpha x}$ along the x propagation direction, where the attenuation constant, α, is given by [2]:

$$\alpha = \left(\frac{2\pi}{\lambda}\right) N_{\text{eff}}^3 \left(\frac{\varepsilon''_{\text{M}}}{\varepsilon'^2_{\text{M}}}\right). \tag{4.26}$$

Consequently, the propagation length is defined as:

$$L_\alpha = \frac{1}{\alpha}. \tag{3.27}$$

The propagation length can be in the order of several micrometers. Contrary to SP waves, optical waveguides have much larger propagation lengths. As the SP waves propagate with high attenuation, they show significant localization of electromagnetic field around the area they are generated. Therefore, the area available for sensing interactions is limited to the region on the metal surface, where the SP waves are excited.

In general, SP waves are generated either by irradiating the thin metal layer's surface with light or by coupling a guided wave onto the boundary of the metal layer and the sample media. The most widely used configurations of SPR sensors are: *prism coupler-based SPR* system (also called *attenuated total reflection—ATR* method); *grating coupler-based SPR* system; and *optical waveguide based SPR system* [3] .

Coupling optical waveguides with the metal layer for generating SP waves has several attractive features for sensing. These include low cost, robustness, relatively small device size, and it provides a simple way for controlling the optical waves' path. An example of coupled optical SP sensing setup using a planar optical waveguide is shown in Fig. 4.12. The optical wave enters the region with a thin metal overlayer, within which the SP waves are established. The sample media containing the measurand is placed over the metallic layer. The SP waves and the guided modes have to be well phase-matched to excite SP waves at the outer interface of the metal.

The ATR configuration, which was first developed by Kretschmann [4] , is widely used in the majority of SPR sensing systems (Fig. 4.13). ATR occurs

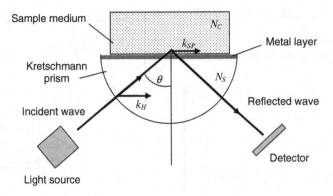

Fig. 4.13 The attenuated total reflectance (ATR) configuration

when light, traveling through an optically transparent medium (with a rather large refractive index of N_S), reaches an interface between this medium and a medium of a lower optical refractive index (N_C). As $N_S > N_C$, total reflection will occur, if the incident angle is larger than the critical angle. The reflected light is detected by an array of photodiodes or charged coupled detectors (CCDs).

The resonance condition of the light in the prism with the SP (coupling into SP) at the metal/sample media interface occurs at a critical incidence angle that depends on the parameters of media. At this resonance angle, the energy from the incident light produces the evanescent SP waves, which consequently reduces the energy of the reflected light.

The wavenumber of the SP, k_{SP}, can be obtained from (4.23) as:

$$k_{SP} = \frac{2\pi}{\lambda c} \frac{1}{\sqrt{N_C^{-2} + \varepsilon'_M{}^{-1}}}. \tag{4.28}$$

The SP resonant occurs, if this wavenumber is equal to the horizontal component of incident wavenumber (k_H), which is given by:

$$k_H = \frac{2\pi}{\lambda c} N_S \sin\theta, \tag{4.29}$$

where N_S is the refractive index of prism and θ is the angle of incident the resonance occurs. The incoming light is able to couple with the SP in the metal at a special angle of incidence corresponding to when $k_H = k_{SP}$, and thus the surface Plasmon is excited. This causes energy from the incident light to be lost to within the metal layer via the lossy propagation of SP waves, resulting in a reduction in the intensity of the reflected light.

An example of the response of an ATR SP sensor after the immobilization of particles such as proteins is shown in Fig. 4.14. The change in the maximum attenuation angle corresponds to the amount of protein attached on the surface. Once proteins come into contact with the surface, rapid adsorption occurs (as N_C changes), which is

Fig. 4.14 A schematic example of protein immobilization and the corresponding response for an SPR sensor

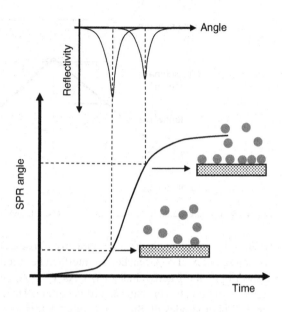

observed as an increase in the angle in which the SPR occurs. This is followed by a plateau in the SPR adsorption, due to saturation of the surface with proteins.

Question 2: A SP-based sensor uses gold as the metallic layer and the system operates at a wavelength of 900 nm. The prism substrate refractive index is $N_S = 2$. Initially the sample refractive index is $N_C = 1.3$, which is that of water, and then is replaced by a material with the refractive index of $N_C = 1.4$. What is the change in the Plasmon resonance angle?

Answer: The Plasmon resonance angle is calculated using:

$$\sin \theta = \frac{1}{N_S \sqrt{N_C^{-2} + \varepsilon'_M{}^{-1}}}. \tag{4.30}$$

If $|\varepsilon'_M| \gg N_C^2$ then the equation can be simplified to:

$$\sin \theta = \frac{\sqrt{\varepsilon'_M N_C^2}}{N_S \sqrt{N_C^2 + \varepsilon'_M}} = \frac{N_C \sqrt{\varepsilon'_M}}{N_S \sqrt{N_C^2 + \varepsilon'_M}} \approx \frac{N_C}{N_S}. \tag{4.31}$$

For gold at 900 nm, $|\varepsilon'_M| = 40$ as can be seen in Fig. 4.15.

For $N_C = 1.3$ or 1.4, the $|\varepsilon'_M| \gg N_C^2$ condition is met. As a result, the resonance angle only depends on the refractive indices of the substrate and the layer.

In this case, $\sin \theta_1 = N_{C1}/N_S = 1.3/2 = 0.65$, as a result $\theta_1 = 40.54°$. Similarly, $\sin \theta_2 = N_{C2}/N_S = 1.4/2 = 0.7$, which results in $\theta_2 = 44.42°$. As a result, the shift in the SPR angle is equal to $\Delta\theta = 3.88°$.

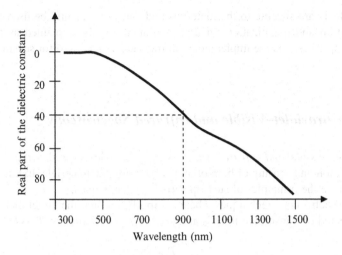

Fig. 4.15 The real part of gold's relative dielectric constant

The sensitivity for the SP resonance sensors to the change in the added layer thickness can also be calculated, similar to that of optical waveguide based sensors. For metals such as gold and silver where $(-\varepsilon'_M \gg N_F^2)$ the sensitivities can be approximated as [2]:

$$\frac{\partial N}{\partial d_{F'}} \approx \frac{2\pi}{\lambda} \left(\frac{1}{N_C^2} - \frac{1}{N_F^2} \right) N^4 \frac{N_C}{(-\varepsilon'_M)^{1/2}}, \tag{4.32}$$

where λ is the optical wavelength. As can be seen, the sensitivity is inversely proportional to wavelength and to $(-\varepsilon'_M)^{1/2}$.

Equation (4.32) describes sensitivity with reference to a layer added on the surface of the sensor. In addition, the change of the effective refractive index is also a function of the change of refractive index of the sample medium. Using (4.32) it can be demonstrated that the sensitivities for SPR sensors are generally 5–10 times larger than those of optical waveguides. Furthermore, by changing the metal thin film and the SPR wavelength, the sensitivity may also be increased. For example, an SPR-based sensor utilizing a gold surface, and having optical wavelength of 632.8 nm, the sensitivity is 1.4 times larger than that of a silver layer device, and it is almost double that of silver at 780 nm [2].

4.4   Spectroscopy Systems

Electromagnetic spectroscopy techniques are used in many sensing applications. As previously presented in Chap. 3, electromagnetic waves of different wavelengths have various effects on the materials, on which they are impinged upon.

These effects are specific to the materials and can consequently be used for their identification and quantization. In this section, it will be explained as how the spectroscopic systems are implemented in transducers and used for sensing.

4.4.1 Ultraviolet–Visible and Infrared Spectroscopy

Ultraviolet–visible and infrared spectroscopy are widely used for quantitative determination and sensing of both organic and inorganic materials. In this section, we introduce the principles of such spectroscopy techniques.

In all spectroscopy techniques a beam is irradiated onto the target material and then reflected or transmitted light is analyzed. The *transmittance*, T, is defined by:

$$T = \frac{P_{out}}{P_0}, \tag{4.33}$$

where P_0 is the intensity of the impinging light on a sample and P_{out} is the intensity of light that emerges from the sample. The *absorbance*, A, is generally defined as:

$$A = \log\left(\frac{P_0}{P_{out}}\right) = -\log T. \tag{4.34}$$

In sensing applications the simplest form of molecular absorption can be related to the concentration of the target materials and can be calculated using *Beer–Lambert law*. The absorbance is directly proportional to the concentration, c (in mol L^{-1}), of the light absorbing species and according to Beer–Lambert law it is obtained by:

$$A = \varepsilon b c, \tag{4.35}$$

where b (cm) is the path length of the material and ε is the *molar absorptivity* or *molar extinction coefficient* (l/mol cm). Molar absorptivity is a characteristic of a substance and indicates the amount of light absorbed at a particular wavelength. The Beer–Lambert law applies to *monochromatic* (single frequency) radiation and is valid for dilute solutions of most substances; however, it fails at high concentrations as solute molecules influence one another as a result of their proximity.

Knowing the absorptivity is an important first step in spectroscopic techniques. By subtracting this effect from the spectra, the remaining signatures, which are observed, will be due to the interactions of the sample materials with the incident light.

A schematic for a basic spectrometer is shown in Fig. 4.16. In this system the light source can be broadband (e.g., produces wavelengths in the range of 400–800 nm for visible spectroscopy). The wavelength selector then filters a single

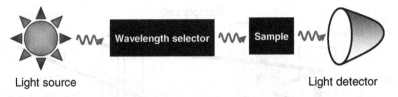

Light source   Light detector

Fig. 4.16 Schematic of a spectrophotometer

frequency beam that passes through the sample. The light detector then produces a signal proportional to the intensity of this monochromic light.

In *spectrophotometry*, several processes occur after the material absorbs energy; these are changes in the *electronic energy*, *vibrational* and *rotational relaxations*, *intersystem crossings*, as well as *internal conversions*. Owing to the quantum nature of molecules, their energy distribution at any given moment can be defined as the sum of the contributing energy terms:

$$E_{total} = E_{electronic} + E_{vibrational-rotational} + E_{translational}. \tag{4.36}$$

The *electronic energy* components correspond to all electrons energy transitions throughout the molecule. For many semiconductors, the electronic transitions can occur when the material absorbs visible and ultraviolet electromagnetic radiations with enough energy for electron transitions. *Vibrational and rotational energies* cause the molecules to undergo a net change in their dipole moments. Vibrational energies are generally higher than the rotational energies and are in the mid-IR region. The nature of such vibrations will be explained in the "Infrared Spectroscopy" section. The energies required for rotational level are quite small (it corresponds to wavelengths of >100 μm in the far IR region). The *translational energies* are related to energies that displace molecules (as opposed to rotational or vibrational).

Having discussed some background regarding electromagnetic spectroscopy, the main techniques that are used in this field, namely, *ultraviolet–visible*, *photoluminescence*, *infrared*, and *Raman spectroscopy* techniques will be presented.

4.4.2   UV–Visible Spectroscopy

Ultraviolet–visible (UV–vis) spectroscopy is widely utilized to quantitatively characterize organic and inorganic materials. A sample is irradiated with electromagnetic waves in the ultraviolet and visible ranges of 200–800 nm (the measurements also sometimes include near infrared range) and the absorbed light is analyzed through the resulting spectrum. It can be employed to identify the constituents of materials, their electronic properties, to determine their concentrations, and to identify some functional groups in molecules. Consequently, UV–vis spectroscopy is used not only for characterization but also for sensing applications. The samples

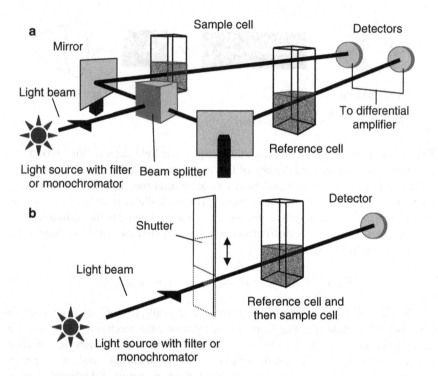

Fig. 4.17 Typical setups for UV–vis spectrophotometers (this example depicts liquid samples—the cells can be replaced by solid- or gas-phase samples): (**a**) double beam configuration and (**b**) single beam configuration

may exist in gaseous, liquid, or solid form. Liquid samples are usually contained in a cell (called a cuvette) that is made of UV and visible light transparent materials such as fused quartz.

Different sized materials can be characterized, ranging from transition metal ions and small molecular weight organic molecules to polymers, supramolecular assemblies, nanoparticles, and bulk materials, whose diameters can be several Ångstroms. Size-dependent properties can be observed in a UV–visible spectrum, particularly in the nano and atomic scales. These include peak broadening and shifts in the absorption wavelength. Many electronic properties, such as the band gap of a material, can also be determined by this technique.

The energies associated with UV–visible ranges are sufficient to excite molecular electrons to higher energy orbitals. Photons in the visible range have wavelengths between 800 and 400 nm, which corresponds to energies between 36 and 72 kcal mol^{-1}. The near UV range includes wavelengths down to 200 nm and has energies as high as 143 kcal mol^{-1}. UV radiation of lower wavelengths is difficult to handle for safety reasons and is rarely used in routine UV–vis spectroscopy.

Figure 4.17a shows typical UV–vis absorption set ups. In the double beam configuration, a beam of monochromatic light is split into two beams, one of them is passed through the sample and the other passes through a reference.

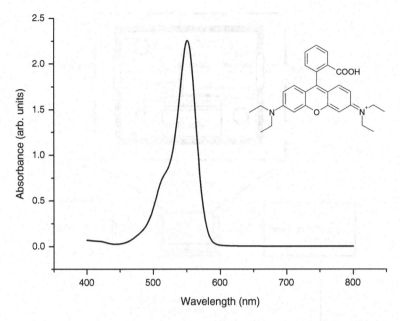

Fig. 4.18 UV–vis absorption spectrum of rhodamine B (shown in *inset*) dissolved in ethanol

After transmission through the sample and reference, the two beams are directed to the detectors, where the output signals are compared. The difference between the signals is the basis of the measurement.

In a single beam configuration, the spectrum of a reference cell is stored and then compared with the spectrum of the sample. Single beam configuration can also be used (Fig. 4.17b). In this case, the reference sample should be scanned first and its spectrum is recorded. Then it is replaced with the target sample. Subtracting the two spectra results in the spectrum of the target sample without any interferences.

An example of a UV–vis absorption spectrum is shown in Fig. 4.18. The spectrum is of rhodamine B, which is a small organic molecule that is commonly used as a dye and as a florescent label for larger molecules. The molecule has a maximum absorbance at around 550 nm. In such measurements, provided that the sample path length and molar extinction coefficient are known, the exact concentration can be determined using the Beer–Lambert law. Additionally, the exact color of the material can also be assessed.

An example of a UV–vis (plus near infra red—NIR) sensing setup for H_2 gas in ambient air, using a metal oxide sensitive layer (WO_3 with platinum (Pt) as a catalyst to enhance the sensing interaction), is presented in Fig. 4.19a [5]. The absorbance vs. optical wavelength for Pt/WO_3 films exposed to 1 % H_2 at room temperature is presented in Fig. 4.19b. Dynamic response of the Pt/WO_3 thin films exposed to different concentrations of H_2 at 100 °C for a single wavelength 660 nm is presented in Fig. 4.19c. For the actual measurements an elevated temperature is used as it increases the H_2 response with the sensitive layer.

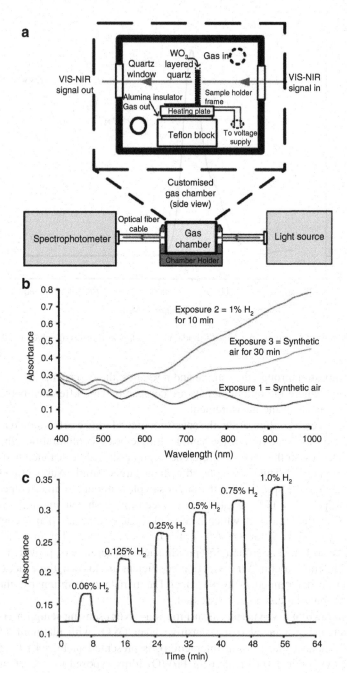

Fig. 4.19 (a) Schematic of the sensing setup. (b) The absorbance versus optical wavelength for Pt/WO$_3$ films exposed to 1 % H$_2$ at room temperature. (c) Dynamic response of the Pt/WO$_3$ thin films exposed to different concentrations of H$_2$ at 100 °C for a single wavelength 660 nm (reprinted from [5] with permission)

4.4.3 Photoluminescence Spectroscopy

Photoluminescence (PL) spectroscopy, for which the fundamentals were presented in Chap. 3, concerns monitoring the light emitted from atoms or molecules after they absorb photons. If these photons excite electrons from the valence band to the conduction band, returning of the electrons to the lower energy band might generate photons with the wavelengths larger than the impinged photons. The photons generated are responsible for the PL effect. The PL spectroscopy is useful for the characterization of both organic and inorganic materials, and the samples can be in solid, liquid, or gaseous forms.

Electromagnetic radiation in the UV and visible ranges is utilized in PL spectroscopy. The sample's PL emission properties are characterized by four parameters: intensity, emission wavelength, bandwidth of the emission peak, and the emission stability. The PL properties of a material can change in different ambient environments or in the presence of other molecules. Many sensors are based on monitoring such changes. As the released photon corresponds to the energy difference between the states, PL spectroscopy can be utilized to study material properties such as band gap, recombination mechanisms, and impurity levels.

As described in Chap. 3, there are several types of PLs and we presented fluoresces and phosphorescence. Fluoresces is the effect which is more frequently used in sensing applications. Fluoresces excitations are generally limited to wavelengths larger than 250 nm as highly energetic shorter wavelengths can cause the rupture of organic bonds. As a result, the fluoresce emissions generally confined to less energetic transitions such as $\pi^* \rightarrow \pi$ (π^* is called pi antibonding orbital). π and π^* are *molecular orbitals*. In chemistry, a molecular orbital is a mathematical function describing the wave-like behavior of an electron in a molecule. A molecular orbital with π-symmetry is a result of either two atomic p_x-orbitals or p_y-orbitals interactions. A π^*-orbital produces a phase change when rotated about the internuclear axis. The most intense fluorescence is found in aromatic compounds that show low energy $\pi^* \rightarrow \pi$ transitions. Florescence can be used in many sensing applications as it changes with temperature, the type of solvent, the pH of the environment, and the concentration of the target analyte.

A typical PL spectroscopy setup is shown in Fig. 4.20. This setup is also called a *fluorometer* for testing fluorescence materials. The source is split into two beams: the first beam passes through a wavelength selector filter or monochromator, then through the sample. The sample generates PL that goes onto a detector. The PL is emitted in all directions. A small portion of the emitted light arrives at the detector after passing through an optional emission filter or monochromator. Generally, a second reference beam is attenuated and compared with the beam from the sample. An emission spectrum is recorded, where the sample is irradiated with a single wavelength, and the intensity of the luminescence emission is assessed as a function of wavelength. The fluorescence of a sample can also be monitored as a function of time, after excitation by a flash of light. This technique is called *time resolved fluorescence spectroscopy*.

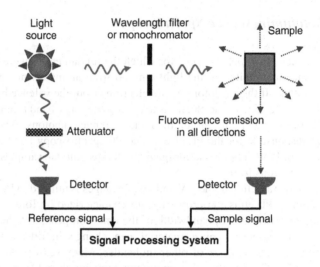

Fig. 4.20 A typical fluorescence spectrophotometer setup

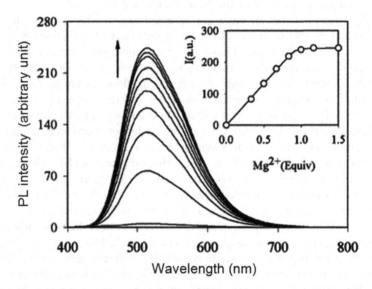

Fig. 4.21 PL emission spectra of the dye molecule (concentration of 5×10^{-5} M) in a methanol–water environment and upon addition of an increasing amount of Mg^{2+} ions from zero to equimolar value concentration to the dye. *Inset*: fluorescence intensity ($\lambda_{ex} = 360$ nm, $\lambda_{em} = 520$ nm) vs. equivalents of Mg^{2+} ions (reprinted from [6] with permission)

Changes in the PL spectra can be used in sensing applications. For instance, Fig. 4.21 shows the PL emission spectra of an organic dye molecule used for monitoring the concentration of magnesium cations (Mg^{2+}) [6]. The PL properties of dye molecules in solution are altered when exposed to the metal cations. With no magnesium ion present, the unexposed molecules exhibit a relatively weak PL emission band at 540 nm. However, the addition of magnesium causes an increase in the emission intensity, which depends on the concentration of added ions (as seen in the inset of the figure) that allows for quantitative detection with a high sensitivity.

4.4.4 Infrared Spectroscopy

Infrared (or *IR*) *spectroscopy* is a popular characterization technique in which a sample is placed in the path of an IR radiation source and its absorption of different IR frequencies is measured. Solid, liquid, and gaseous samples can all be characterized using this technique.

Near IR (NIR) regions associate with the wavelengths in the range of 0.8–2.5 μm. The most important application of this region is the sensing of industrial and agricultural materials. Species such as water, low molecular weight hydrocarbons, food, petroleum, and chemical industries have signatures in this region.

Most used IR spectrum has the wavelength in the range of 2.5–25 μm (commonly shown as 4,000 to 400 cm^{-1}). The frequency scale at the IR charts is commonly given in units of reciprocal centimeters (cm^{-1}) rather than Hz, because the numbers are more manageable (e.g., 1,000 cm^{-1} is the magnitude for the wavelength of 5 μm) and also the number is proportional to the energy of the wavelength). These wavelengths are insufficient to excite electrons to higher electronic energy states, but are sufficient to generate transitions in vibrational energy states. These vibrational states are associated with molecular vibrations, and consequently each molecule has its own unique signatures. These states are relevant to the configuration of molecular bonds. Therefore, IR spectroscopy may be employed to identify the type of bond between two or more atoms and consequently identify functional groups. Because IR spectroscopy is quantitative, the numbers of a type of bond in a system can be determined.

Virtually all organic compounds absorb IR radiation, but inorganic materials are less commonly characterized, as heavy atoms show vibrational transitions in the far IR region, with some having extremely broad peaks that make the identification of the functional groups difficult. Furthermore, the peak intensities of some ionic inorganic compounds may be too weak to be measured.

The covalent bonds that hold molecules together are neither stiff nor rigid, but rather they vibrate at specific frequencies corresponding to their vibrational energy levels. The vibration frequencies depend on several factors including bond strength and the atomic mass. As an analogy, the bonds can behave in a similar manner to springs which are connected to masses. As shown in Fig. 4.22, chemical bonds may be contorted in six different ways: stretching (both symmetrical and asymmetrical), scissoring, rocking, wagging, and twisting. Absorption of IR radiation causes the bond to move from the lowest vibrational state to the next highest, and the energy associated with absorbed IR radiation is converted into these types of motions.

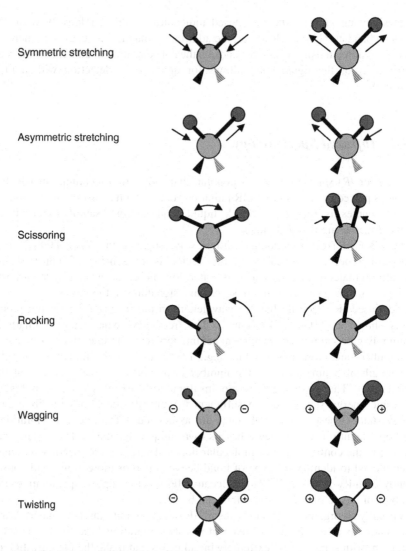

Fig. 4.22 Six different ways in which molecular bonds can vibrate

Far infrared include the 25–1,000 μm range that has lower energy and can be used for rotational spectroscopy.

The IR peaks of a simple molecule such as carbon dioxide (CO_2) gas molecules can be readily analyzed. CO_2 molecules are linear and hence have four fundamental vibrations (Fig. 4.23a). The asymmetrical stretch of CO_2 produces a strong band at ~2,350 cm^{-1}. The two scissoring or bending vibrations are equivalent and therefore, have the same frequency and are said to be *degenerate*. The IR signature for these vibrational modes appears at ~650 cm^{-1}. The IR spectrum of the CO_2 can be seen in Fig. 4.23b.

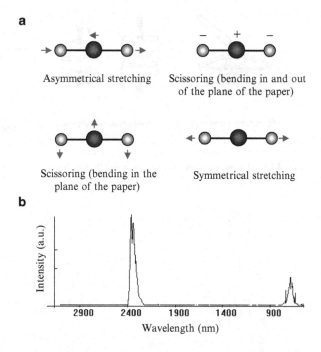

Fig. 4.23 (a) Stretching and bending vibrational modes for CO_2 and (b) the FTIR spectrum for CO_2 gas

Complex molecules contain dozens or even hundreds of different possible bond stretches and bending motions, which implies the spectrum may contain dozens or hundreds of absorption lines. This means that the IR absorption spectrum can be its unique *fingerprint* for identification of a molecule. A *diatomic molecule*, that has only one bond can only vibrate in one direction. For a linear molecule (e.g., hydrocarbons) with n atoms, there are $3n - 5$ vibrational modes. If the molecule is nonlinear (such as methane, aromatics, etc.), then there will be $3n - 6$ modes.

Samples can be prepared in several ways for an IR measurement. For powders, a small amount of the sample is added to *potassium bromide (KBr)*, after which this mixture is ground into a fine powder and subsequently compressed into a small, thin, quasi-transparent disk (Fig. 4.24). For liquids, a drop of sample may be sandwiched between two salt plates, such as NaCl. KBr and NaCl are chosen as neither compounds shows an IR-active stretch in the region typically observed for organic and some inorganic molecules. Acquired IR spectra are subtracted from a background spectrum to remove unwanted signals, in particular, any water present in the ambient.

The IR transmittance of the small organic dye molecule 9-anthracene carboxylic acid is shown in Fig. 4.25. The peaks can be assigned to different vibrational motions of specific bonds. For example, aromatic compounds like anthracene show stretching of the C–H bonds at wavenumbers between 3,130 and 3,070 cm^{-1}, and several in-plane C–H bending peaks between 1,225 and 950 cm^{-1}. Peaks observed between 1,615 and 1,450 cm^{-1} are attributed to both stretching in the aromatic ring, all of which can be seen in the spectrum.

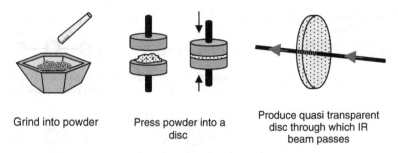

Grind into powder   Press powder into a   Produce quasi transparent
                                  disc   disc through which IR
                                                               beam passes

Fig. 4.24 The preparation of a KBr disk for IR spectroscopy

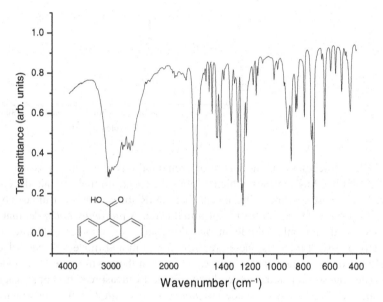

Fig. 4.25 An IR spectrum of an organic molecule (9-anthracene carboxylic acid, shown in the *inset*)

These days the miniaturized IR systems are commercially found for a wide range of applications. One of such sensing systems is referred to as *non-dispersive IR* (*NDIR*) systems and is commonly used for gas sensing (Fig. 4.26). The main component of such systems is an IR source, a sample chamber, a wavelength filter, and an IR detector. The target gas is sent into the sample chamber, and the absorption of a specific wavelength in the IR region is observed by the detector. The optical filter eliminates all light except the wavelength that the target gas molecules and ideally other gas molecules in the environment should not absorb light at this wavelength. Generally, there is also another chamber with a reference gas (typically nitrogen).

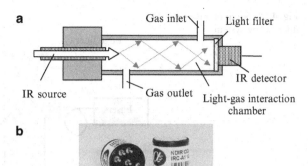

Fig. 4.26 (a) Schematic of a NDIR gas sensor, (b) the photo of an NDIR sensor courtesy of Aplhasense Ltd.

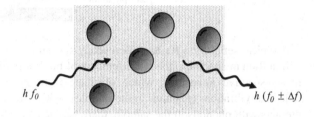

Fig. 4.27 Representation of Raman scattering from particles

The issue of cross sensitivity needs to be addressed in NDIR systems, as many gases have common IR signatures. For instance, H_2O, which always exists as the water vapor in the ambient gas, has a broad peak in the $3,500$–$2,700$ cm^{-1} region.

4.4.5 Raman Spectroscopy

Raman spectroscopy is based on monitoring the intensity and wavelength of light that is scattered inelastically (that means the scattered light shows a shift in wavelength) from the sample. Raman spectroscopy was first discovered in 1928 by C.V. Raman, an Indian Scientist, who received the 1931 Nobel Prize for this discovery. It is suitable for characterizing organic and inorganic samples and generally considered a complementary technique to IR spectroscopy.

In Raman spectroscopy, a sample is irradiated with light of known polarization and wavelength (generally in the visible or infrared ranges). Inelastic (or Raman) scattering occurs and the scattered light is wavelength shifted with respect to the incident light (Fig. 4.27). The spectrum of the scattered light is then analyzed to determine the changes in its wavelength. Raman spectroscopy is a powerful analytical tool for qualitatively and quantitatively investigating the composition of materials. Raman spectrum studies are also ideal for sensing and characterizing target analytes in liquid environments. Unlike IR spectroscopy, where water signal interference can block out entire regions of the spectrum, a Raman spectrum is less susceptible to the presence of water.

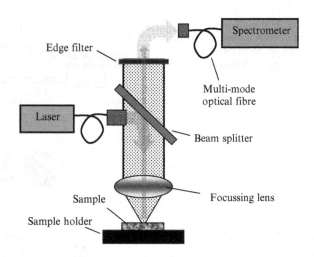

Fig. 4.28 Setup of a Raman microscopy system

A typical setup for a Raman microscopy system is shown in Fig. 4.28.

In a Raman spectrum, the wavenumbers of the Raman shifts are plotted against their respective intensities, which originate from the interaction photons with molecular vibrations (phonons in the sample). When irradiation from a laser source interacts with phonons in the sample, there can be an exchange of energy between them. The phonons may either gain or lose energy. The phonon modes are intrinsic properties related to chemical bonding and therefore, the information contained within a Raman spectrum may provide a *fingerprint* by which molecules can be identified. This not only serves for characterization purposes. Raman spectra can also be used in sensing applications. In sensing applications, monitoring the peak intensities and positions can provide quantitative information on the number of analyte molecules that have taken part in the interaction, such as in affinity-based sensing.

If the Raman scattered photon has a lower energy than the incident light, its frequency is shifted down and is referred to as a *Stokes emission.* On the other hand, if the scattered photon has higher energy than the incident light, the frequency is shifted up and it is referred to as an *anti-Stokes emission.* The energy of the scattered photon, E, is related to the energy of the incident photon, $E_0 = hf_0$, by:

$$E = hf_0 \pm \Delta E_v, \tag{4.37}$$

$$\text{Stokes} : f = f_0 - \Delta f, \tag{4.38}$$

$$\text{Anti-Stokes} : f = f_0 + \Delta f. \tag{4.39}$$

where ΔE_v is the change in energy and Δf is the change in frequency.

Raman systems should be designed and implemented with care and accuracy. Raman scattering comprises a very small fraction of the scattered light, only one Raman scattered photon from 10^6 to 10^8 excitation photons. Therefore, the main limitation of Raman scattering is detecting the weak inelastically scattered light from the intense Rayleigh scattered light. Modern instrumentation almost universally employs notch or edge filters for blocking the signal from the excitation laser. Furthermore, depending on the incident light energy, photoluminescence may occur, which can obscure the Raman spectrum.

The Raman scattering intensity is inversely proportional to the fourth power of the emission wavelength. So decreasing the wavelength of the light source should result in an increase in the Raman signal. However, decreasing the wavelength increases the likelihood of observing photoluminescence. Stokes shifts are susceptible to photoluminescence interference, but not anti-Stokes.

The Raman signal can be enhanced if molecules are adsorbed on roughened metal surfaces, typically gold or silver. This is called *surface enhanced Raman spectroscopy (SERS)*, and exploits changes in analyte polarizability perpendicular to the surface, which enhances scattering by factors of more than a million-fold. The metal surface must have a Plasmon in the frequency region close to that of the excitation laser. Surface roughness or curvature is generally required for the generation of strong surface Plasmons. In order for SERS to occur, the particles or features must be small when compared the wavelength of the incident light. SERS-active systems should possess structures typically in the 5–100 nm range. SERS is an ideal technique in sensors, particularly for monitoring very small traces of an analyte.

Figure 4.29 shows an example of SERS in chemical sensing [7]. Assembly of silver nanowires that with ~50 nm diameter and 2–3 μm length are used to form the sensitive layer. A fluorescent molecule, Rhodamine 6G (R6G), which produces a distinct Raman spectrum, when excited by light of wavelength 532 nm, is adsorbed onto the films. The inset of the figure shows that there is a linear relationship between the R6G concentration and the intensity of the Raman peak at $1,650 \, \text{cm}^{-1}$. This system was capable of detecting 0.7 pg of the analyte, which is an impressive detection limit.

4.4.6 Nuclear Magnetic Resonance Spectroscopy

Nuclear magnetic resonance (NMR) spectroscopy is a complex technique employed largely to study the chemical structure of organic and inorganic compounds, either in liquid or solid form. Nuclei of atoms that contain odd numbers of protons or neutrons (such as hydrogen-1 and carbon-13) have an intrinsic magnetic moment, due to their nuclear spin. These spinning charges create magnetic moments. As a result of the exposure to a strong magnetic field, the magnetic moments can line up either with or against the field. This occurs when the magnets spin clockwise or counter clockwise. The difference of energy between the two states depends on the strength of the external magnetic field and even at the strength of several Tesla is very small ($\sim 0.1 \, \text{cal/mol}^{-1}$), which is equal to the

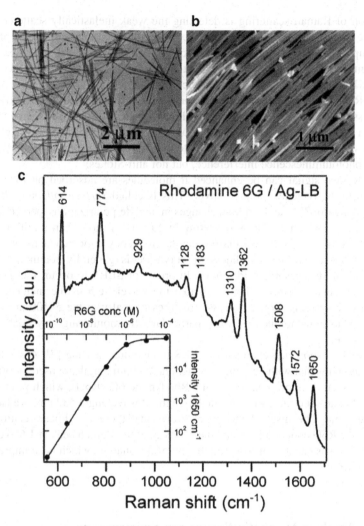

Fig. 4.29 (a) The transmission electron microscopy micrograph of the silver nanowires, (b) the scanning electron microscope micrograph of the wires, when orderly aligned on the substrate, and (c) SERS spectrum of R6G on the film (incident light: 532 nm, 25 mW) after 10 min incubation in a 10^{-9} M R6G solution. The *inset* shows the calibration curve. It is the relationship between the Raman intensity at 1,650 cm^{-1} and the R6G concentration (reprinted from [7] with permission)

electromagnetic frequency energy in the MHz to low GHz range. Consequently, the spin alignment can be perturbed using radio frequency (RF) electromagnetic fields. The resulting response to the perturbing field is used in NMR spectroscopy. NMR spectroscopy is increasing becoming more important in sensing, particularly for the detection of compounds and for determining the chemical structure of polymers and biomolecules.

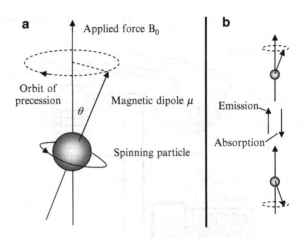

Fig. 4.30 (a) Effect of magnetic field on a rotating particle. (b) Model for the absorption and emission of a precessing particle

A classical description of the NMR process is as follows: assume that a particle with a magnetic field is spinning along its axis as shown in Fig. 4.30. If a magnetic field of B_0 is applied, due to the gyroscopic effect, the axis of the rotating particle starts precessing around the vector of the magnetic field. The angular velocity of this motion is given by:

$$\omega_0 = \gamma B_0. \tag{4.40}$$

in which γ is the *magnetogyric ratio* and is calculated using $\mu = \gamma p$, where p is the angular momentum and μ is the magnetic moment of the charged particle. The potential energy of the precessing charged particle is then:

$$E = -\mu B_0 \cos \theta. \tag{4.41}$$

for which θ is the angle between the applied field vector and magnetic moment of the particle. When an RF energy is applied, it is absorbed by the particle and this angle of precession changes (perturbs). We assume that the absorption involves flipping of the magnetic dipole in the opposite direction as shown in Fig. 4.30. In reality, as the process occurs in the quantum level only certain magnetic dipoles are allowed. For the flipping to occur, there must be a magnetic force applied at the right angle to the B_0 fixed field. This magnetic force must oscillate in tune with the precession of the particle. If this condition applies, flipping occurs.

When the RF field is turned off, the system loses its energy and transforms back into its initial state. This process can be through the spin–lattice or spin–spin relaxation. Generally, the signal decay has a first-order trend in the order of 0.1 s.

If the electromagnetic radiation has the right frequency, the material's nucleus resonates, or flips, from one magnetic alignment to another, known as the resonance condition. Because the resonant frequencies depend not only on the nature of the atomic nuclei but also on the atoms, which are bound to that nucleus and chemical environment, functional groups and chemical bonds can be identified.

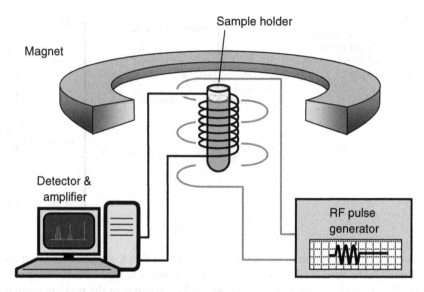

Fig. 4.31 Schematic diagram of an NMR spectroscopy setup

An example of a typical NMR spectroscopy setup for liquid samples is shown in Fig. 4.31. The dissolved sample in a solution is placed between the poles of a large magnet. This solution contains a small amount of material, such as *tetramethylsilane* (TMS—$Si(CH_3)_4$), which produces a standard absorption line in the acquired spectrum. A pulse of RF radiation (in the order of several hundred MHz for 1–10 μ s) causes nuclei in the magnetic field to flip to a higher energy alignment. The nuclei then re-emit RF radiation at their respective resonant frequencies (this process needs time in the order of 0.1–1 s), which creates an interference pattern in the received signal. The emitted RF radiation is collected with a sensing coil. The frequencies of the RF emissions are extracted through Fourier transformation.

An NMR spectrum shows a plot of *chemical shift* against the intensity of absorption. Since both the frequency shift and fundamental resonant frequency are directly proportional to the strength of the magnetic field, the shift is converted into a field-independent, dimensionless, value known as the chemical shift. The chemical shift is reported as a relative measure from the reference resonance frequency (for the nuclei ^1H and ^{13}C, TMS is commonly used as a reference). The chemical shift is defined as the difference between the frequency of the signal and the frequency of the reference divided by frequency of the reference signal to give:

$$\delta = \frac{\text{Observed chemical shift from the reference peak (e.g. TMS)}}{\text{Spectrometer frequency}}. \quad (4.42)$$

Delta, δ, is the relative shift which is generally scaled to 1 ppm (part per million) of the NMR's operating frequency. The absolute values of frequency shifts are

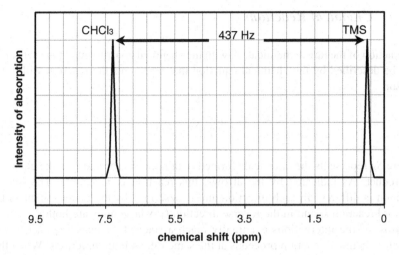

Fig. 4.32 Proton NMR spectrum of CHCl₃

extremely small in comparison to the fundamental NMR frequency. A typical frequency shift might be in the order of several 10 s of Hz. NMR charts are calibrated using an arbitrary scale called the delta scale. To illustrate this, an NMR spectrum of chloroform ($CHCl_3$) tuned for the nucleus of hydrogen atoms, operating at 60 MHz, is shown in Fig. 4.32. An NMR measurement tuned for the hydrogen nucleus is called proton, or ^1H, NMR. The $CHCl_3$ peak shown is measured relative to the TMS peak. If the RF frequency on the NMR is set to 60 MHz, then $CHCl_3$ produces a chemical shift of 437 Hz/60 MHz, which is equal to 7.28 ppm. For molecules with more intricate structures (or those not containing hydrogen at all) the ^1H-NMR may not provide sufficient information. In such cases, the NMR instrument may be tuned for the nuclei of isotopes such as ^{13}C, ^{15}N, ^{19}F, ^{29}Si, and ^{31}P.

One of the major applications of NMR spectroscopy systems is for the structural study of complicated organic biomolecules, where it can provide detailed information on their 3D structures. Large complex molecules such as proteins, DNA/RNA, typically exhibit several thousand resonances and among them are many inevitable overlaps. As a consequence, complicated mathematical techniques employing multidimensional NMR experiments are required, and the assignment of peaks is carried out with advanced computational methods.

4.5   Electrochemical Transducers

Electrochemical transducers generate signals that result from the presence and interaction of chemical species. They make use of various chemical effects to monitor concentrations of these species. As will be demonstrated, using these effects, we study and quantify the concentration of target analytes and monitor chemical reactions.

4.5.1 Chemical Reactions

Interaction of the target chemical, X, with the chemical constituents of the sensor, S, can be described by the chemical equation that represents the reaction within the sensor:

$$X + S \underset{k_r}{\overset{k_f}{\rightleftharpoons}} S_X + R, \tag{4.43}$$

where S_X represents the chemicals formed within the sensor and R is the chemical byproduct. As indicated by the arrows, this reaction is reversible. The rate of reaction is different in each direction and is described by the rate constant in the forward reaction k_f and in the reverse direction k_r whose units are both s^{-1}.

Most of the interactions eventually reach a state of *chemical equilibrium* in which a chemical reaction proceeds at the same rate in both directions. When this condition is met, there is no change in the concentrations of the various compounds involved. This process is known as *dynamic equilibrium*. Without any energy input, chemical reactions always proceed towards equilibrium. For a reaction of the type:

$$aA + bB \rightleftharpoons cC + dD, \tag{4.44}$$

the *reaction quotient*, Q_P, provides an indication of whether the reaction shifts to the right or left and is very commonly approximated as:

$$Q_P = \frac{[C_i]^c [D_i]^d}{[A_i]^a [B_i]^b}, \tag{4.45}$$

where i denotes the instantaneous concentration at a moment in time and a, b, c, and d moles of reactants [A], [B], [C], and [D] participate in this reaction. The mole is a unit of measurement used in chemistry to express amounts of a chemical substance, defined as a value of 6.02×10^{23} elementary entities (e.g., ions, molecules, atoms) of that substance.

The equation is more accurately described, if the molar concentrations are replaced by the activity of a species defined by $a_X = \gamma_X[X]$, in which γ_X is the *activity coefficient*. γ_X describes how actively the species can influence the equilibrium. For not very active species (when their ionic strength is minimal) and when the solution is not concentrated, ions are sufficiently apart from each other, so they do not to influence one another's behavior. As a result, the species affect the equilibrium independent of their type and only based on their molar concentration. In this case, γ_X is unity and does not appear in (4.45) and Q_p only depends on the molar concentration of species.

When reaction (4.44) is in a state of equilibrium, the concentrations of reactants and products are related by the *equilibrium constant*, K:

$$K = \frac{[C]^c [D]^d}{[A]^a [B]^b}. \tag{4.46}$$

It is important to note that non-interacting liquids, solvents, and solids are not included in these equations, as their concentrations remain constant.

4.5.2   Chemical Thermodynamics

A chemical sensor's performance strongly depends on the energies that are released or absorbed, during the chemical reactions. Many chemical reactions have a tendency to proceed in a direction in which they are *energetically favorable* or *spontaneous*. Therefore, by knowing the spontaneity of a reaction, an indication of the likelihood of the reaction occurring can be obtained. This is important in electrochemical-based sensing, as it provides an indication of the response dynamics of a sensitive material towards an analyte species.

In a spontaneous chemical reaction, energy is released from the system, causing it to become thermodynamically more stable. The spontaneity can be deduced by determining the change in free energy available for doing work, which is called the *Gibbs free energy*:

$$\Delta G = \Delta H - T\Delta S, \tag{4.47}$$

where ΔH is the change of the system's *enthalpy*, which at constant pressure is the same as the heat added or removed from the system, T is the temperature, and ΔS is the change in the systems *entropy*.

Enthalpy is a measure of the total energy of a thermodynamic system. It includes the internal energy and the amount of energy that is required for displacing it (pressure, which produces a system volume). It is described as $H = U + pV$, in which U is the internal energy, p is the pressure, and V is the volume of the system. Enthalpy is a thermodynamic potential and has the unit of joule.

From statistical calculations, entropy is obtained as a logarithmic measure of Ω which is the number of microstates in a system:

$$\Delta S = k_B \ln \Omega, \tag{4.48}$$

where k_B is the Boltzmann constant, 1.38×10^{-23} J/K^{-1}.

The free energy conditions for spontaneity are:

$\Delta G < 0$, spontaneous (favored) reaction,

$\Delta G = 0$, system in equilibrium, no driving force prevails,   (4.49)

$\Delta G > 0$, non-spontaneous (disfavored) reaction.

From the *second law of thermodynamics*, entropy of a chemical reaction that is not in equilibrium tends to increase. This increase reduces the order of the initial system, and therefore entropy is an expression of disorder or randomness. In a thermodynamic system, pressure, density, and temperature tend to become uniform over time because this equilibrium state has higher probability (more possible combinations of microstates) than any other. For instance when an ice cube melts in a glass full of water, the difference in temperature between a warmer room (the surroundings) and cold glass of ice and water (the system and not part of the room) begins to be equalized. The thermal energy from the warm surroundings spreads to the cooler system of ice and water. Eventually, the temperature of the glass and its contents and the temperature of the room become equal. In this case the entropy of the room has decreased as some of its energy has been transferred to the glass of water and ice. Conversely, the entropy of the system of ice and water has increased. This increase is more than the decrease in the entropy of the surrounding room.

From (4.47), to ensure the reaction is spontaneous and hence $\Delta G < 0$, it is seen that change in the enthalpy, ΔH, must be sufficiently negative. If the free energy of the reactants in a chemical reaction occurring at constant temperature and pressure is higher than that of the products, $G_{reactants} > G_{products}$, then the reaction will occur spontaneously.

The Gibbs free energy, at any stage of the reaction, can be found through its relationship with the reaction quotient in (4.45), which is calculated as [8]:

$$\Delta G = \Delta G^0 + RT \ln(Q_P), \tag{4.50}$$

where ΔG^0 is the standard-state free energy of reaction and R is the gas constant $(8.314472 \, \text{J K}^{-1} \, \text{mol}^{-1})$. Equation (4.50) is extracted from (4.48) in which Q_p is the direct description of the microstates Ω and $N_A R = k_z$ in which N_A is the *Avogadro constant*. When the reaction reaches equilibrium, $\Delta G = 0$ and the reaction quotient takes on the value of the equilibrium constant from (4.46), the change in free energy at equilibrium becomes:

$$\Delta G^0 = -RT \ln(K) \tag{4.51}$$

As will be demonstrated later, (4.50) and (4.51) are of fundamental importance to the function of electrochemical sensors. K, *equilibrium constant*, is a function of analyte concentrations. It will be seen that ΔG^0 is a function of voltage produced by the electrochemical interactions. As a result, the concentrations of target analytes can be obtained using voltage or current measurements.

4.5.3 Nernst Equation

In this section, the fundamental basis of *electrochemical sensors* will be presented. In such sensors, electrochemical measurements are used for analytical assessments.

In fact, largest and oldest group of chemical sensors are electrochemical devices. Sensors as diverse and popular as blood glucose monitors and high temperature metal oxide gas sensors in automobiles are all included in this category.

Electrochemistry deals with the transfer of charge from an electrode to its surrounding environment. During an electrochemical process, chemical changes take place at the electrodes and charges transfer through the bulk of the sample. Electrochemistry is based on *redox (reduction–oxidation) reactions*. A redox reaction involves transfer of electron from one species to another. A species is oxidized when it loses electrons. It is reduced when it gains electrons. An *oxidizing agent*, which is also called an *oxidant*, receives electrons from another substance and is reduced in the process. A *reducing agent*, which is also called a *reductant*, donates electrons to another substance and *oxidizes*.

For understanding the performance of electrochemical sensors, we first need to become familiar with some basic concepts in electrochemistry including: *Galvanic cells, reference electrodes, salt bridges*, and *standard reduction potentials*.

A *Galvanic* (or *voltaic*) *cell* uses spontaneous chemical reactions to generate electricity. In such a cell, one reagent oxidizes and another reduces and a voltage difference is produced as a result of these reactions. If an electrode is placed in an *electrolyte solution* (an electrolyte solution is a substance that dissociates into free ions when dissolved, to produce an electrically conductive medium), it generates a potential.

It is important to consider that for many interactions the net reaction is spontaneous but no little current flows through an external circuit. This generates no flow of current through the external circuit to be measured for assessing the reaction. For instance in:

$$Cd(s) + 2Ag^+(aq) \rightleftharpoons Cd^{2+}(aq) + 2Ag(s), \tag{4.52}$$

aqueous Ag^+ ions can interact directly at the $Cd(s)$ surface, which generates no net charge and hence no current. (s) denotes solid and (aq) aqueous conditions. This means that such interactions cannot be monitored in a sensing measurement. In order to avoid this problem, we can separate the reactant into two half-cells and connecting these two half-cells using a *salt bridge* or via a *membrane*. Each of the electrode–electrolyte system is called a *half-cell*. Always a combination of two electrode–electrolyte system is needed to generate a voltage.

A *standard reduction potential* (denoted by $E°$) can be used to predict the generated voltage when different half-cells are connected to each other. The term standard means that the activities of all species are unity. A *hydrogen electrode* is generally used as the standard reference. Such a reference electrode consists of a Pt surface in contact with an acidic solution for which $A_{H+} = 1$ mol. A stream of H_2 gas (1 bar pressure and 25°C) is purged through the electrode to saturate the electrode with aqueous H_2. The reaction is:

$$H^+(aq) + e^- \rightleftharpoons \frac{1}{2}H_2. \tag{4.53}$$

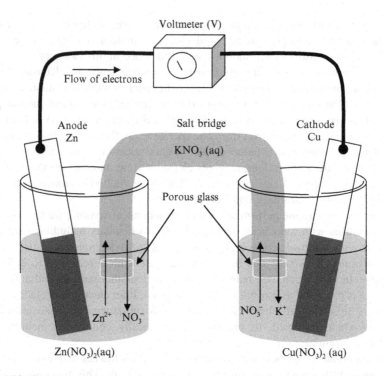

Fig. 4.33 The electromotive force generated in an electrochemical cell as a result of the charge flow

A potential of zero is assigned to this *standard hydrogen electrode* (*SHE*).

In a Galvanic cell, the generated voltage is the difference between electrode potentials of the two half-cells. The magnitude of this potential depends on: (a) the nature of electrodes, (b) the nature and concentrations of solutions, and (c) the liquid junction potential at the membrane (or the salt bridge). An example of a conventional Galvanic cell setup, with zinc and copper electrodes, is shown in Fig. 4.33. If concentrations of electrolytes are both 1 mol, then the potential is measured to be equal to +1.1 V, as:

$$Zn^{2+} + 2e^- \rightarrow Zn(s) + 0.763\,V, \tag{4.54}$$

$$Cu^{2+} + 2e^- \rightarrow Cu(s) - 0.337\,V, \tag{4.55}$$

$$Zn(s) + Cu^{2+} \rightarrow Zn^{2+} + Cu(s) + 1.100\,V. \tag{4.56}$$

The Gibbs free energy for this reaction is negative, showing that the reaction proceeds spontaneously at room temperature. This cell can be practically employed as a battery. Some standard electrode potentials are listed in Table 4.1.

Table 4.1 Standard
electrode potentials

Reaction	E^0 (V) at 25°C
$Pb^{4+} + 2e^- \rightleftharpoons Pb^{2+}$	+1.695
$O_2(g) + 4H^+ + 4e^- \rightleftharpoons 2H_2O$	+1.229
$Ag^+ + e^- \rightleftharpoons Ag(s)$	+0.799
$Fe^{3+} + e^- \rightleftharpoons Fe^{2+}$	+0.771
$AgCl(s) + e^- \rightleftharpoons Ag(s) + Cl^-$	+0.222
$2H^+ + 2e^- \rightleftharpoons H_2$	0
$Cd^{2+} + 2e^- \rightleftharpoons Cd(s)$	−0.403
$Zn^{2+} + 2e^- \rightleftharpoons Zn(s)$	−0.763
$Ti^{2+} + 2e^- \rightleftharpoons Ti(s)$	−1.634
$Li^+ + e^- \rightleftharpoons Li(s)$	−3.096

In Fig. 4.33 the salt bridge which consists of a tube filled with a high concentration of KNO_3 (or KCl) is shown. This provides the electrical contact between the two cells, while avoid the mixing of two electrolytes in the compartments. The ends of the bridge are covered with porous glass disks that allow ions to diffuse but minimize the mixing of solutions inside and outside the bridge. In this case, K^+ from the bridge migrates into the cathode compartment and a small amount of NO_3^- migrates from the cathode into the bridge. Similarly Zn^{2+} migrates to the anode compartment, while $2NO_3^-$ migrates in reverse. Ion immigration offsets the charge build-up that would occur in the bridge so the bridge produces minimal voltage.

The electrode can only donate or accept electrons from the media, which is immediately adjacent to their surface. Interestingly this media can have a composition that is starkly different from that of the bulk of the electrolytes. This area is generally considered to have a bilayer formation, which is referred to as the *electrical double layer* (Fig. 4.34). The first layer of molecules on the surface of the electrode is adsorbed by van der Waals forces. The next layer is established when ions in the electrolyte are attracted by the electrode charges. The electrode charges can be due to the applied voltage to the electrode or induced by the charges of electrolyte in its vicinity. This region, in which the composition is different from the bulk solution, is called the *diffuse part* of the double layer and can be from a few nanometers to a few micrometers thick depending on the concentration of ions and the applied or formed voltages. Any given solution has one *potential of zero charge* (POZC) at which there is no excess charge on the electrode. The POZC can be obtained by applying an external voltage to the electrodes and changing it, while observing the current–voltage (*I–V*) curve.

Electrical work carried out by an electrochemical cell equals the product of the charge flowing and the potential difference across. If we operate the electrochemical liquid cell at a constant pressure and temperature cell, then the work carried out in the cell is:

$$W = -E \times q, \tag{4.57}$$

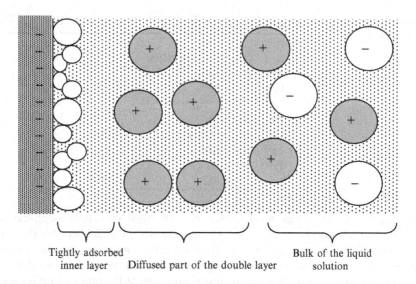

Tightly adsorbed   Bulk of the liquid
inner layer   Diffused part of the double layer   solution

Fig. 4.34 The electrical double layer formed on the electrode surface

where E is the electromotive force (emf) of the cell in Volts (this parameter is denoted using V in the rest of the book but E is chosen here to be consistent with the conventional texts in electrochemistry literature) and q is the charge flowing through the cell, which is calculated from:

$$q = n \times N_A \times e, \tag{4.58}$$

where n is the number of moles of electrons transferred per mole of reaction, N_A is Avogadro's number ($6.02 \times 10^{+23}$), and e is the charge of an electron (-1.6×10^{-19} C).

F is generally defined as $N_A \times e = F$ (F is the Faraday constant, which is equal to 96,487 C mol^{-1}), thus:

$$W = -nFE. \tag{4.59}$$

The free energy change for a chemical reaction conducted reversibly at constant temperature and pressure equals the electrical work that can be carried out by the reaction on its surroundings:

$$W = \Delta G. \tag{4.60}$$

As a result, the Gibbs free energy relates to the voltage of the cell through:

$$\Delta G = -nFE. \tag{4.61}$$

From (4.50) [considering that $\ln(K) = 2.303 \log(K)$] and defining $\Delta G^0 = -nFE^0$, we can obtain:

$$E = E^0 - 2.303 \left(\frac{RT}{nF}\right) \log(K) \qquad (4.62)$$

or

$$E = E^0 \frac{-[0.05916(V)]}{n \log(K)}, \qquad (4.63)$$

which is called the *Nernst equation*, which describes the cell potential. The Nernst equation is the base of all electrochemical sensor measurements. The equation has been named after Walther Nernst, who was a Nobel Prize recipient of German origin in 1920.

Example 1. What is the potential for a half-cell containing a cadmium electrode in a solution of Cd^{2+} ions of 0.01 M concentration?

Answer: The stranded potential of this reaction is presented in Table 4.1 as:

$$Cd^{2+} + 2e^- \rightarrow Cd(s) \quad E^0 = -0.403 \text{ V}.$$

From (4.63), we can write ($n = 2$ as two electrons participate):

$$E = E^0 \frac{-[0.0591(V)]}{2 \log(1/[Cd^{2+}])},$$

which results in:

$$E = \frac{-0.403 \text{ V} - [0.0591(V)]}{2 \log(1/0.01)} = -0.4621 \text{ V}.$$

In sensing applications, metallic electrodes are used extensively. Their choice depends on the cost and the activity of the target analyte ions. The metallic electrodes are commonly categorized into *first*, *second*, and *third kinds* and *inert redox* electrodes.

The electrodes of the *first kind* are pure metals in equilibrium with their cations in an electrolyte. For instance, for copper:

$$Cu^{2+}(aq) + 2e^- \rightleftharpoons Cu(s). \qquad (4.64)$$

Such electrodes are not selective and respond to many other cations. For instance, silver cations also interact with Cu electrodes. Many such electrodes can be easily dissolved in acids. Additionally, they can be easily oxidized.

A *second kind* electrode is made of a metal that can become responsive to an anion with which it can form a stable complex. Silver–silver chloride reference electrode is of this type, which will be discussed in the next session.

The *third kind* electrodes consist of metals that under special circumstances respond to other cations. The mercury electrodes can be of this type.

Metallic redox types are electrodes made from inert metals such as gold, platinum, and palladium. In this case the electrode can act both as the source or sink of electrons without participating in the interaction. The electron transfer at such electrodes can be irreversible so for slow reactions, they cannot produce reproducible and predictable results.

4.5.4 Reference Electrodes

It is common to utilize a *reference electrode* in sensing processes in electrochemical cells. This electrode does not participate in interactions and its properties are known, so its effect can be easily removed from the measured signal. In practice for sensing applications, the reference electrodes are implemented which are easy to setup, are non-polarizable, and give reproducible electrode potential, which have low coefficients of variation with temperature. Many varieties of such electrodes are available but two of the most common ones are: *silver–silver chloride* and the *saturated calomel* electrodes.

Silver chloride is not soluble in water. Consequently, it can be used in many aqueous sensing applications without affecting the target analyte. In a two-cell configuration, the half-cell reaction of the silver–silver chloride reference electrode is given as follows:

$$AgCl(s) + e^- \rightarrow Ag(s) + Cl^- \quad E^0 = +0.22 \, \text{V}. \tag{4.65}$$

Consider the example cell shown in Fig. 4.35. In this example, the system is set up for measuring the relative concentrations of Fe^{2+} and Fe^{3+}. Pt is used as the working electrode as it does not interact with the aqueous Fe ions. The two half-cell reactions can be written as follows:

$$Fe^{3+} + e^- \rightleftharpoons Fe^{2+} \quad E^0 = +0.77 \, \text{V}, \tag{4.66}$$

$$AgCl + e^- \rightleftharpoons Ag(s) + Cl^-(aq) \quad E^0 = +0.22 \, \text{V}. \tag{4.67}$$

The electrode potentials are then obtained as ($n = 1$ for both interactions):

$$E_+ = +0.77 \, \text{V} - [0.0591 \, (\text{V})] \times \log\left(\frac{[Fe^{2+}]}{[Fe^{3+}]}\right), \tag{4.68}$$

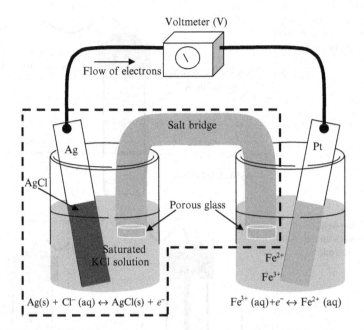

Fig. 4.35 The half-cell within the *doted space* is called a silver–silver chloride reference electrode

$$E_- = +0.22\,\text{V} - [0.0591\,(\text{V})] \times \log([\text{Cl}^-]). \qquad (4.69)$$

The concentration of Cl^- is constant, which is maintained by the highly saturated KCl. Therefore, the measured differential voltage $E = E_- - E_+$ only changes when ratio of $[\text{Fe}^{2+}]/[\text{Fe}^{3+}]$ changes. As a result, this system can be reliably used as a Fe ion sensor.

A commercial silver–silver chloride reference electrode generally consists of a silver wire or silver substrate coated with silver chloride that is dipped in a solution of a salt such as potassium chloride (either saturated KCl or 3.5 M). The deposition of the silver chloride layer is a commercially viable practice. For instance, it can be obtained by making the silver plate as the anode of an electrochemical cell with a platinum cathode and potassium chloride as electrode. Several minutes of electrolyzation with a positive potential, as small as 0.5 V, oxidizes the silver surface to silver ions. The silver ions attract chloride ions to form a silver chloride film in the process.

The voltage of a reference electrode with saturated KCl is approximately +0.197 V. Figure 4.36 shows an Ag/AgCl reference electrode setup.

The saturated calomel electrode (mercury chloride) is another common reference electrode in electrochemical sensors. The half-cell reaction is:

$$\text{Hg}_2\text{Cl}_2 + 2e^- \rightarrow 2\text{Hg} + 2\text{Cl}^- \quad E^0 = +0.268\,\text{V}. \qquad (4.70)$$

In practice, the electrode voltage in saturated KCl drops slightly and is approximately +0.241 V.

Fig. 4.36 Schematic setup of a commercial silver–silver chloride reference electrode

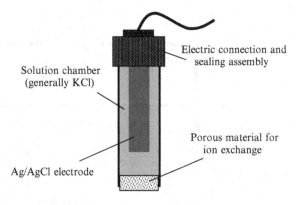

Electric connection and sealing assembly

Solution chamber (generally KCl)

Porous material for ion exchange

Ag/AgCl electrode

Fig. 4.37 Ion-selective membrane with an external reference electrode

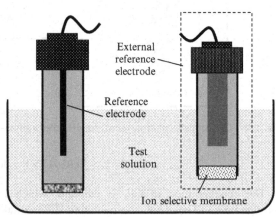

External reference electrode

Reference electrode

Test solution

Ion selective membrane

4.5.5 Membrane Electrodes

Membrane electrodes or *ion selective electrodes* (*ISEs*) are electrodes that by incorporating membranes in their structure can respond selectively to target ions in the presence of many other ions. They are employed to measure specific ions in a solution or in a gas phase. These sensors are generally made of membrane-based electrochemical setups. The membrane is generally a component that makes the electrode selective to a particular ion. The membrane is somehow the replacement for the salt bridge.

These ion-selective membranes are fundamentally different from metal electrodes as they do not involve in the redox process themselves. A voltage difference develops across the membrane (due to the generation of the junction potentials), when it is placed in a solution. To measure this voltage, this ion-selective membrane is used in combination with an internal or external reference electrode (Fig. 4.37). The relationship between the measured potential, E, and the ion activity in the sample is also mathematically described by the Nernst equation, as this voltage is generated at a junction.

Membrane should have the following qualities: (a) minimal solubility in the sensing environment, (b) conductivity to ions, and (c) selectivity to the target ions.

Electrochemical tests and processes now represent a large number of hospital procedures and ISEs play an important role in them. For instance, blood chemistry tests are often ordered prior to a surgery or a medical procedure to assess the general health of a patient. This blood test, commonly referred to as *Chem 7 tests*, examines seven different substances found in the blood. They include: blood urea nitrogen (BUN), carbon dioxide (CO_2), creatinine, glucose, serum chloride (Cl^-), serum potassium (K^+), and serum sodium (Na^+). These days, most of the components in Chem 7 tests are analyzed with ISEs. No electrode responds selectively to only one kind of ion. For instance, the glass pH electrode is among the most selective to hydrogen ions. However, these membranes are also, to some degree, responsive to sodium and potassium ions as an interfering species.

The ISEs can be categorized into different types, depending on the material of the membrane:

Glass membrane: It is used for measuring ions such as Na^+ or measuring pH. Glass membrane electrodes are formed by doping the silicon dioxide glass matrix with various chemicals. The most common glass membrane electrodes are the pH electrodes. *Corning* 015 is the most widely used glass membrane for the sensing of H^+ ions (pH measurements) with approximately 22 % Na_2O, 6 % CaO, and 72 % SiO_2. Commonly called *chalcogenide glass*es are other type of glass membranes that have selectivity for double-charged metal ions, such as Pb^{2+} and Cd^{2+}.

Crystalline membranes: Crystalline membranes are made from mono- or poly-crystals of ionic compounds (e.g., NaCl is an ionic compound). In these membranes only ions, which can introduce themselves into the crystal structure, affect the electrode response. As a result, they generally show good selectivity to such ions. An example is the fluoride selective electrode based on LaF_3 crystals that has an excellent selectivity to fluoride ions (F^-).

Polymeric membranes: Polymer membrane electrodes consist of various ion-exchange materials incorporated into an inert matrix of polymer such as *polyethylene (PE)*, *polyvinyl chloride (PVC)*, *polyurethane (PU)*, and *polydimethylsiloxane (PDMS* or commonly called silicone). Polymeric membranes are the most widespread electrodes with ionic selectivity to analytes such as potassium and calcium ions. They are also used for measuring ions including fluoroborate, nitrate, and perchlorate. However, such electrodes have generally low chemical and physical durability.

Liquid-membrane electrodes: They are formed from immiscible liquid that selectively bonds to certain ions. They are the most useful membranes for sensing polyvalent cations.

Gas permeable membrane: Sensing electrodes are available for the measurement of gas species such as ammonia, carbon dioxide, dissolved oxygen, nitrogen oxide, sulfur dioxide, and chlorine. These electrodes have a gas permeable membrane. Many metal oxides such as *zirconia* are of this type.

4.5.6   An Example: Electrochemical pH Sensor

In many industrial, chemical, and medical processes, pH is an important parameter to be measured and controlled. The pH of a solution indicates how acidic or basic (alkaline) it is. As described in the previous section, the glass-based materials is the most commonly used membranes in pH sensors. When measuring pH, we measure the negative log of hydrogen activity as:

$$pH = -\log[H^+{}_{activity}], \tag{4.71}$$

$$[H^+{}_{activity}] = 10^{-pH}. \tag{4.72}$$

The conventional pH readings range is from 0 to 14. Using an electrochemical pH sensor, the potential develops across the membrane is measured, when the systems are in contact with a solution. The Nernst equation can be used for the calculation of pH according to (4.53) as:

$$E = E^0 - [0.05916\,(\text{V})] \times \log\left(\frac{1}{[H^+]}\right) = E^0 + [0.05916\,(\text{V})] \times \log([H^+]),$$

$$\tag{4.73}$$

or

$$E = E^0 - 0.05916\,pH. \tag{4.74}$$

As can be seen, the output voltage changes linearly with the pH of the environment. One pH unit corresponds to 59.16 mV at 25 °C, which are the standard voltage and temperature at which all calibrations are referenced. Most of the electrochemical pH sensors are also sensitive to temperature, so its effect should be accounted for in measurements.

Apart from industrial applications, pH sensors are useful in biosensing applications. In many of such sensors, enzymes are used in the structure of electrodes. Several types of enzymes are able to produce H^+ or OH^- ions, when they interact with specific target biomolecules, and their concentration can then be estimated using pH sensors.

4.5.7   An Example: Electrochemical-Based Gas Sensor

Zirconium oxide (zirconia) electrochemical oxygen sensor is one of the most common gas sensors. In such a sensor, the electrolyte is typically made of zirconia (Fig. 4.38). The sensor electrodes can be made of platinum, which also operates as a catalyst for splitting the target gas and generating ions that can pass thorough the material.

Fig. 4.38 A schematic
of zirconia oxygen sensor

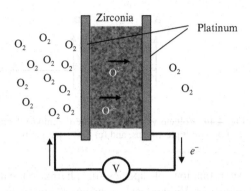

Fig. 4.39 Schematic of an
industrial zirconia gas sensor
incorporated into the exhaust
of an engine

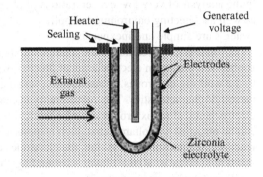

A zirconia-based gas sensor operates at elevated temperatures of higher than 400 °C, at a temperature above which zirconia becomes an ion conductive material.

Due to the oxygen gas partial pressure difference on each side of zirconia, the migration of oxygen ions from a higher concentration to a lower concentration side occurs. The diffusion of oxygen ions produces a voltage across the device, which is proportional to the gas pressure difference between the two sides. Schematic of a commercial oxygen sensor installed in a gas exhaust is shown in Fig. 4.39. The whole system is also governed by the Nernst equation.

4.5.8   Voltammetry

Voltammetry is widely used for assessing the adsorption processes at the surface of electrodes and electron transfer mechanism. It was initially developed in 1920s and chemists were using it extensively for sensing of inorganic ions. In mid-1960s classical voltammetry was further modified and the access to electronic circuits enhanced the sensitivity and selectivity of the method. The emergence of low cost and high gain *operational amplifiers* was a significant contributor. Nowadays voltammetry techniques are used comprehensively in the sensing and the

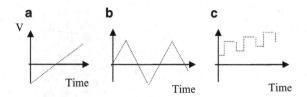

Fig. 4.40 Voltage vs. time for some common schemes of voltammetry. (**a**) CV—one sweep, (**b**) CV—multiple sweeps, and (**c**) *SWV* on a ramp

determination of species in pharmaceuticals, environmental, and biological industries. Voltammetric sensors are finding increasing use in medical applications, in the analysis of very low concentrations of pharmaceuticals and their metabolites as well as detecting environmental pollutants. Generally, electroanalytical sensing systems are simple and inexpensive.

In voltammetry techniques the relationship between currents and voltages (*I–V*) is observed during an electrochemical process, when a variable potential is applied to the working electrodes. Consequently, sensing information can be derived from these *I–V* characteristics. The voltammetry systems can provide information about electrochemical redox processes and chemical reactions. Since transient responses can be obtained by voltammetry, such responses can be used for studying very fast reaction mechanisms. In addition, the electrodes can be used as tools for producing reactive species in a small layer surrounding their surfaces to monitor chemical reactions involving target species.

Linear sweep voltammetry (LSV—also called polarography), *cyclic voltammetry* (*CV*), and *square wave voltammetry* (*SWV*) are the most widely used signaling schemes in voltammetric techniques (the applied input voltage signal for each of these types is demonstrated in Fig. 4.40.

The schematic of a voltammetric setup (*voltagram*) is shown in Fig. 4.41. It consists of a working electrode, where redox interactions occur and a reference electrode that provides a closed loop for the flow of current. Very commonly a third electrode, made of materials such as platinum, is also implemented in the system as a counter electrode. In this case, the voltage is applied between the counter and reference electrodes, and the current and voltage are measured between the working and reference electrodes. The electrolyte solution in the cell generally contains an *indifferent electrolyte* (or *supporting electrolyte*) along with an *oxidizable* or *reducible species* (*electroactive species*). The indifferent electrolyte does not participate in the interaction and only makes the solution conductive.

When the power supply forces electrons into and out of the system, the charged surface of the working electrode attracts ions of opposite charge. The charged electrode and the oppositely charged ions next to it form an *electric double layer* as described previously (Fig. 4.34).

Commonly in voltammetry the signal from the sources is fed into the counter electrode. The reference electrode voltage is measured using a very large input impedance circuit that draws no current. Consequently, the current is mainly

Fig. 4.41 (**a**) A voltammetric sensing system and (**b**) depiction of electrodes in a sample holder: W, R, and A stand for working, reference, and counter electrodes, respectively

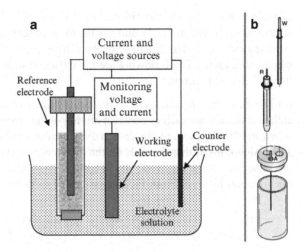

sourced from the counter electrode to the working electrode. As a result, the independent voltage for measure is the voltage difference between the working and reference electrodes.

The working electrode can take different shapes and forms. The most common ones are conductors deposited onto inert substrates. The most frequent conductors are noble metals such as gold and platinum; carbon materials such as graphite and carbon nanotubes; metal coated with mercury and also conductive transparent materials such as indium tin oxide. In addition to metallic electrodes, a large number of other types of electrodes are investigated and implemented in sensing applications to achieve the qualities required in sensors. In such cases, the electrode materials are chosen in order to provide qualities including sensitivity and selectivity. The electrodes can also be modified to provide opportunities for other applications such as *electrochromic* effects in *smart windows*.

In voltammetry, various range of potentials can be used which depend on the electrode material and the composition of electrolyte. One of the limitations is caused by the oxidation of water producing oxygen at large currents in large positive voltages. Similar effect occurs at negative voltages, when hydrogen gas is produced.

The electrochemical process and the shape of *I–V* characteristics, which is obtained during voltammetry, depend on *diffusion* and *capacitive currents*, which are as follows:

4.6   Diffusion Current

The diffusion current can be calculated from the *Fick's first law of diffusion*:

$$J = -D\left(\frac{\partial c}{\partial x}\right),\tag{4.75}$$

where J is the *diffusion flux* (in mol m^{-2} s) per unit area per unit time, D is the *diffusion coefficient* (m^2 s^{-1}), and c is the local concentration of chemical specie in volume (mol m^{-3}). The behavior of the diffusion current depends on the environmental conditions. The following are the most common conditions in the observation of diffusion current:

(a) *The case that the thickness of the double layer remains constant*: This can be archived via actions such as constant mixing of the electrolyte and when the system is in a stable condition. In this case, a dynamically stable system is established with a diffusion current that depends on the difference of target analyte concentrations on the surface of the electrode and in bulk of the solution as well as the thickness of the diffusion layer, δ. From Fick's first law of diffusion:

$$\frac{(dN / dt)}{A} = -D\left(\frac{(c_a - c_s)}{\delta}\right), \tag{4.76}$$

where N is the number of moles of the target analyte, A is the electrode surface area, and c_s and c_a are the concentrations of analyte near the surface and in the bulk of analyte, respectively. Subsequently, the diffusion current can be calculated from the general *Faradiac relationship*:

$$i(t) = \left(\frac{dN}{dt}\right)nF, \tag{4.77}$$

where n is the number of electrons take part in the redox interaction and F is the Faraday number.

Equation (4.76) is only valid for a constant potential. By changing the applied voltage, and before reaching the *potential of zero charge* (POZC that was discussed previously) , the difference between the two concentrations increases. As the solution is constantly mixed, the bulk analyte concentration remains constant. However, the concentration of ions on the surface of electrodes, c_s, is altered. The *diffusion limiting current* occurs when the voltage at the surface is so large that all ions exchange electrons, and consequently, an ion-free surface meaning $c_s = 0$ is generated. If we combine (4.76) and (4.77), the diffusion limiting current will be equal to:

$$i_d = \frac{nFADc_a}{\delta}, \tag{4.78}$$

Obviously, this current is proportional to the value of c_a. As a result, it can be a key parameter in sensing for the determination of the concentration of the target analyte. In the process of increasing the applied voltage, i_d is the maximum value for the measured current. When current is one half of the value of the limiting current, $i = i_d/2$, the voltage is called the *half-wave potential*, $E_{1/2}$. The typical *I–V* characteristic of a constantly mixed system is shown in Fig. 4.42. As can be seen,

Fig. 4.42 The I–V characteristic of an electrochemical cell, when the system is well mixed

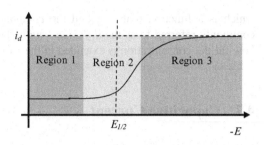

the I–V characteristic in region 2 is similar to a diode's I–V characteristic. However, it tapers and eventually reaching saturation in region 3.

(b) *The concentration of the analyte is low and the mixing is not aggressive*: In this case, the concentration of the analytes in the bulk, c_a, media change and the thickness of the diffusion layer alters with voltage changing. Therefore, the *Fick's second law of diffusion* is used (which can be extracted from the Fick's first law together with the *mass conservation law* as $\frac{\partial c}{\partial t} = -\frac{\partial J}{\partial x}$):

$$\frac{\partial c}{\partial t} = D\left(\frac{\partial^2 c}{\partial x^2}\right). \tag{4.79}$$

The analyte concentration gradient, $\frac{\partial c}{\partial x}$, and hence $\left(\frac{\partial^2 c}{\partial x^2}\right)$, is affected by the combination of: the diffusion layer thickness change and the concentration change. They can change in different directions, with different rates, and the effect of one may dominate the other one.

A commonly observed example is when a charge is placed on the surface of electrode (like an applied pulse) and then the system is allowed to become stable in time. This is generally referred to as homogenization when the system eventually becomes uniform. This means the total concentration of the ionic species (charge) is constant as $\int_0^\infty c\,dx = B$. In this case, the concentration can be obtained as:

$$c(x,t) = \frac{B}{\sqrt{4\pi Dt}} c^{-\frac{x^2}{4Dt}}. \tag{4.80}$$

This function has a bell shape distribution that tapers in time.

Another commonly observed example is when the concentration of the surface c_s remains constant and the concentration of the bulk can change, the solution of (4.79) will become:

$$1 - \frac{2}{\sqrt{\pi}} \int_0^{x/2\sqrt{Dt}} e^{-z^2}\,dz, \tag{4.81}$$

which is a function that is called the *complementary error function*. $2\sqrt{Dt}$ is commonly referred to as the *diffusion length*. This number provides a measure of how far the concentration is extended in the x direction.

4.7 Capacitive Current

In addition to the diffusion current, capacitive currents also exist in electrochemical processes. The double layer forms a capacitive dielectric region between the bulk analyte area and the surface of the electrode. This capacitance is known as the *double layer capacitance* and its value is proportional to the surface area of plates. The capacitive current is essentially unwanted in most of the sensing applications as the sensing information is generally extracted from the diffusion current curves.

In addition to the capacitive current, there are also interferences such as adsorption currents caused by adsorption/desorption of molecules on electrodes. The most common interferent molecules are surfactants.

4.7.1 Chronoamperometry (or Potential Step Voltammetry)

The electrochemical system is generally rather complicated. However in voltammetry, it is a good approximation to assume that the charge on the electrode remains constant for a small time duration and use (4.81) to predict the current behavior. This results in the *Cortel equation* as:

$$i(t) = \frac{nFAc_a\sqrt{D}}{\sqrt{\pi t}},\qquad(4.82)$$

This equation is especially important in *chronoamperometry* (or *potential step voltammetry*) when a constant voltage is applied. This means that current drops off with at a rate proportional to $(\pi t)^{-1/2} \times D^{1/2}$. As a result, knowing the diffusion constant of the species the concentration of the target analyte can be obtained (Fig. 4.43).

4.8 Linear Sweep Voltammetry

LSV is a common electrochemical voltammetric sensing technique. Although, many sophisticated voltammetric techniques were developed to replace LSV addressing its deficiencies, its relative simplicity still makes it an attractive choice for electrochemical sensing systems.

Fig. 4.43 Experiments
showing how the Cortell
equation describes current as
a function of time in a
chronoamperometry
measurement

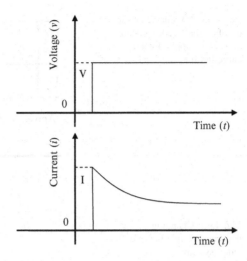

Fig. 4.44 *I–V* characteristic
of an electrochemical system
when the system's rate
constant is large

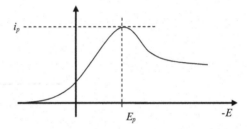

In slowly changing charge condition on electrodes (voltage change at a low or moderate rates), the current magnitude will eventually reach a maximum value at E_p with its corresponding i_p. The change of current as a function of voltage, in a typical LSV, is shown in Fig. 4.44. The current initially increases as a result of an increase in the difference of analyte concentration and the surface concentration. However, after the peak, when voltage increases further, current decreases as a result of the increase of the diffusion layer thickness. The number of the peaks can be more than one depends on the type of surface reactions.

In LSV, the applied voltage is linearly scanned. The rate of voltage change can be in the order of 0.01–100 mV s^{-1}. However, it depends on the concentration of the target analyte as well as materials and dimensions of electrodes. The characteristics of the LSV depend on three major factors: (1) the rate of the electron transfer, (2) the chemical reactivity of the species, (3) and the voltage scan rate.

In voltammetric experiments, the current response is generally plotted as a function of voltage. Take for example the Fe^{3+}/Fe^{2+} redox system:

$$Fe^{3+} + e^{-} \rightleftharpoons Fe^{2+}. \tag{4.83}$$

Fig. 4.45 *I–V* characteristics of a Fe^{3+}/Fe^{2+} redox system. *Inset* is the applied voltage curve in time

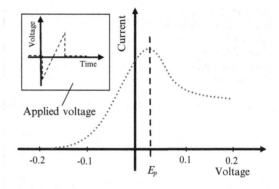

Fig. 4.46 Changing the scan rate in a Fe^{3+}/Fe^{2+} redox system

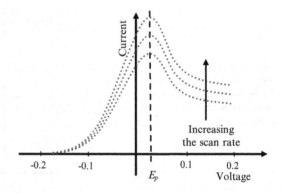

The single voltage scan voltammogram of this system is presented in Fig. 4.45. Such systems are important in the release of Fe ions in Ferritin protein within our body regulating the body's iron storage. A similar electrochemical process is used for measuring the Fe ion concentrations of our body.

In a Fe^{3+}/Fe^{2+} redox system, for voltages smaller than -0.2 V no current is seen. As the voltage increases, the current increases and reaches a peak value of E_p. Further increase in the voltage decreases the current, as the diffusion layer thickness increases.

The *sweep rate* is an important factor in the *I–V* curve behavior. It determines how fast the electrons can diffuse into the double layer and interact at the surface of the electrodes with ions. The reversibility of the process is also determined by the sweep rate. Figure 4.46 shows the change in the linear sweep voltammograms as the scan rate is increased. As it can be observed, the curves remain proportionally similar. However, the current increases with the increasing scan rate. This current increase can be readily attributed to the change in the diffusion layer thickness.

The value of the peak currents is proportional to the square root of the scan rate, $i_p \propto \sqrt{u_t}$. In slow voltage scans, the diffusion layer grows further from the electrode surface than in fast scans. Consequently, the ion flux to the electrode surface becomes smaller decreasing the current magnitude.

Fig. 4.47 Change in the current magnitude of increasing TNT concentrations (in 250 ppb steps) (reprinted from [9] with permission)

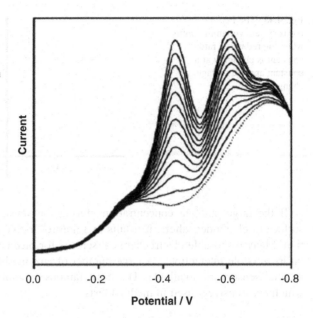

A rapid system, which can exchange ions and electrons in a much shorter time than the scan rate, is generally a *reversible electron transfer system*. Conversely, for a slow electron transfer system, the *I–V* characteristics depict *quasi-reversible* or *irreversible electron transfer systems*. In reversible systems, a repeating cycle gives the same response and reversing the voltage sweep produces a mirrored *I–V* curve. Electrodes in reversible reactions can be reused as their surface does not change.

An example for the response of an electrochemical sensor in a LSV measurement setup is shown in Fig. 4.47. It is for the detection of TNT in liquid environment [9]. This sensor assembly consisted of the carbon-fiber, silver–silver chloride, and platinum working, reference, and counter electrodes, respectively. The figure shows voltammograms for seawater samples containing increasing levels of TNT in 250 parts per billion (ppb) steps. As can be seen, the two cathodic peaks show increase in the current magnitude due to the presence of larger concentrations of TNT in each step.

In Fig. 4.48 the *I–V* curves for the Fe^{3+}/Fe^{2+} redox system voltammograms are recorded as the constant reduction rate (k_{red}) is changing at a constant applied voltage change rate.

Decreasing the rate constant decreases the concentrations of ions at the electrode surface and slows down the reaction kinetics. In this case, the equilibrium is not established rapidly. As a result, the position of the current peak shifts to the higher voltages upon the reduction of the rate constant. However, decreasing the rate constant makes the system less reversible.

Generally charge transfer reactions are reversible if $k > 0.1$–1 cm s^{-1} and irreversible for $k < 10^{-4}$ to 10^{-5} cm s^{-1}. They are referred as quasi-reversible for values that fall between reversible and irreversible.

Fig. 4.48 The Fe^{3+}/Fe^{2+} redox system voltammograms when the reduction rate constant is changing at a constant applied voltage change rate

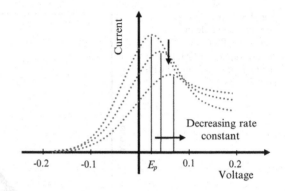

If the target analyte concentration, during the sensing measurements or the surface of electrodes, alters, it results in a non-reversible electrochemical interaction. Many disposable electrochemical sensors for medical applications are based on irreversible interactions. A large number of enzyme-based voltammetric commercial sensors are available. The most famous example is the glucose sensor, which is extensively used in medical tests.

4.8.1 Cyclic Voltammetry

Cyclic voltammetry is similar to LSV, with a difference that the potential applied to a working electrode is frequently altered in time (a triangular waveform as shown Fig. 4.40). The voltage change rate is generally in the range of 0.1–10,000 mV s^{-1} for electrodes with the area of approximately 1 mm^2. This voltage repeatedly oxidizes and reduces the species located within the diffusion layer near the surface of electrode.

The emergence of powerful data acquisition systems, measurement equipment, and associated microelectrodes has made it possible to increase the voltage change rate and measure extremely small currents. As a result, it is now possible to identify species that exist for just a few micro- and nanoseconds, and even measure individual electron transfer reactions. The I–V characteristics also provide a powerful means to study the redox reaction energies to investigate the dynamics and reversibility of the electron transfer as well as the rates of coupled chemical reactions. These advantages have ensured that the voltammetry systems have been widely adopted for studying biochemical/chemical reactions, environmental sensing, and the monitoring of industrial chemical components.

Figure 4.49 shows a typical cyclic highly reversible voltammogram. When the voltage is altered, before the onset of the E_i voltage (the voltage for overcoming the ionic junction barrier), no measurable current is observed. At this point, the voltage of the working electrode reaches the threshold value causing the reduction of the target species. This generates a current, associated with the reduction process.

Fig. 4.49 A cyclic
voltammogram of a
reversible, one-electron (only
one cathodic and one anodic
peaks) redox reaction

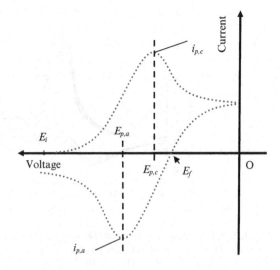

Following this, the current increases rapidly as the concentration of near surface
free ions decreases, where the diffusion current reaches a peak at $E_{p,c}$ (cathodic
peak). As the potential decreases further, the thickness of the diffusion layer
increases. This results in decay of the current. In this example the final voltage is
centered at 0 V.

In this example, when the voltage reaches 0 V, its polarity alternates and
increases again. However, the value of current is kept constant and a reduction
reaction proceeds at the electrode's surface, which is caused by the residual charges
within the diffusion layer. This cathodic current continues to decrease. At E_f, the
overall number of oxidation interactions becomes equal to the number of the
reduced species adjacent to the electrode surface and the current will become
zero. Increasing the voltage further depletes the reduced material at the surface of
electrode. It reaches a minimum current correspond to a voltage peak of $E_{p,a}$
(anodic peak). As the potential returns to E_i, the magnitude of current decreases
as the thickness of the diffusion layer increases.

In a near ideal reversible electrode reaction, the difference in peak potentials
expected to be equal to:

$$\Delta E_p \approx \left| E_{p,a} - E_{p,c} \right| = \frac{0.05916}{n}, \tag{4.84}$$

in which n is the number of electrons involved. If the system is irreversible then
ΔE_p exceeds the value calculated by this equation. Fortunately, the electron transfer
is almost reversible for most of the slow rate reactions. Generally, by increasing the
sweep rate ΔE_p value increases. As a result, the unset of irreversibility can be used
for assessing the rate constant. Cyclic voltammetry can also be utilized to interpret
complex behavior of electrochemical interactions. The nature of the double layer
changes, if several oxidation reduction reactions occur during each cycle.

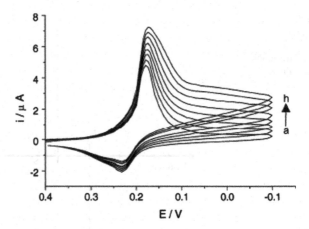

Fig. 4.50 Cyclic voltammograms of the working electrode in a phosphate buffer (pH 6.8) (**a**) solution without any cholesterol and with cholesterol concentration of (**b**) 1×10^{-6}, (**c**) 2×10^{-6}, (**d**) 3×10^{-6}, (**e**) 4×10^{-6}, (**f**) 5×10^{-6}, (**g**) 5×10^{-6}, (**h**) 5×10^{-6} mol L^{-1}. Scan rate 50 mV s^{-1} (reprinted from [10] with permission)

Example in Fig. 4.50 shows the response of a cholesterol biosensor fabricated by the immobilization of cholesterol oxidase (which is an enzyme. Enzymes and their functions will be presented in Chap. 5.) in a layer of silica matrix [10]. The system is able to sense cholesterol in the concentration range of 1×10^{-6} to 7×10^{-6} mol L^{-1}. Increase in the ionic species increases the cathodic peak value as expected.

4.8.2   An Example: Stripping Analysis

In *stripping analysis*, analyte from a diluted solution is adsorbed into a thin film of Hg (mercury, which is a metal in liquid form) in the electrochemical reaction. The electroactive species is then striped from the electrode by reversing the direction of the voltage sweep. Current measured during the oxidative removal is proportional to the analyte's concentration. Stripping is one of the most sensitive methods for the sensing of *heavy metal ions*.

4.9   Solid State Transducers

Solid state transducers are made of *metal–semiconductor* and/or *semiconductor–semiconductor junctions* and work by monitoring changes in the device's *electrical field distribution* in the presence of the measurand. Parameters that can be directly measured include voltage, current, capacitance, and impedance, and from them, a

Fig. 4.51 (a) Schematic of a p–n junction (**b** and **c**) typical *I–V* curves of a p–n junction based sensor: (**b**) lateral changes of the curve and (**c**) total downward shift of the curve

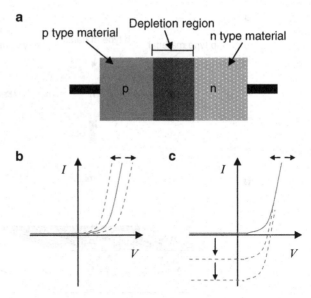

myriad of electrical properties (such as electrical conductivity, barrier height, and carrier concentrations) can be derived. The popularity of solid state transducers is built upon well-established microfabrication strategies, originating from the silicon microelectronics industry. Semiconductor-based devices such as diodes and transistors are naturally sensitive to environmental changes. In their performance, the numbers of free electrons and holes in the transducer structure change and the electric field distribution are altered in response to external stimuli.

In this section, some of the most popular solid state transducers are introduced and their application in sensing will be presented.

4.9.1   p–n Diodes and Bipolar Junction Based Transducers

The *p–n diodes* or *bipolar junction* (*BJ*) devices are based on semiconductor–semiconductor junctions. When implemented in sensing applications, electrical properties, such as barrier height and carrier concentration, can be altered due to the presence of a measurand. These changes result in a change in relationships between current, voltage, and accumulated charge.

p–n junction devices are based on semiconductors, which are doped, so that the materials have a majority of free electrons (n-type) or free holes (p-type). When p-type and n-type semiconductors are placed next to each other to form a p–n junction, the majority carries in the p-type material (holes) diffuse a certain distance into the n-type material, and similarly at the n-type material, majority carries (electrons) diffuse into the p-type material. Once this diffusion is balanced, a depletion region forms (Fig. 4.51a). A potential barrier then establishes between the

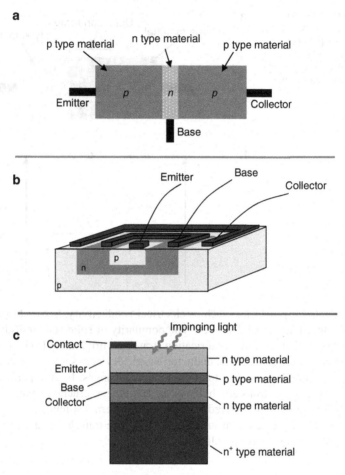

Fig. 4.52 (a) A schematic of a BJT, (b) cross-sectional illustration of a homojunction BJT, (c) schematic of a heterojunction phototransistor

p- and n-doped materials, which must be exceeded for the current to conduct. For instance, if this voltage is 0.7 V, an applied voltage of larger than this magnitude will be needed to observe the passage of any current through the junction. Devices having one p–n junction are termed diodes, whilst those contained two, in the form of p–n–p or n–p–n junctions (Fig. 4.52a), are referred to as *bipolar junction transistors (BJTs)*.   BJTs have the added advantage of internal current amplification.

Using the *Shockley equation*, the current flow through a p–n junction as a function of voltage, V, across the device is given by [11]:

$$I(V) = I_{\text{saturation}}(e^{qV/nkT} - 1), \tag{4.85}$$

where q is the electron charge, n is the ideality factor, k is the Boltzmann's constant, T is the temperature in Kelvin, and $I_{\text{saturation}}$ is the saturation current of the device. The saturation current can be obtained using:

$$I_{\text{saturation}} = SA^{**}T^2 \exp(-q\phi_b/kT), \qquad (4.86)$$

where S is the area of the metal contact (cm^2), A^{**} is the effective *Richardson's constant* ($A \ \text{cm}^{-2} \ \text{K}^{-2}$), and ϕ_b is the *barrier height*.

The I–V curves of a typical p–n junction sensor and how they shift upon exposure to a measurand are demonstrated in Fig. 4.51b, c. The curve can show lateral shift, if the measurand affects the properties of the junction. Change of barrier height and temperature are such effects (Fig. 4.51b).

Devices such as diodes and BJTs are also commonly employed for monitoring charge and temperature. They may also be employed, albeit less commonly, for chemical and pressure sensing applications. However, the most extensive usage of such transducers is in light sensing and irradiation spectroscopy. When a photon of sufficient energy strikes the diode in the depletion region, it creates a free electron–hole pair that can be swept from the junction by the built-in field as was described in Chap. 3. Thus a photocurrent is produced and the I–V curve shifts downwards (Fig. 4.51a).

In many spectroscopic measurements, the amount of light to be monitored may be very low. Using a photodiode in conjunction with an amplifier may rectify such limitations. However, low irradiation intensity produces small currents, which are inherently difficult to amplify externally, if electrical noise is to be avoided. In such situations, phototransistor may be used as an alternative to external signal amplification.

A BJT has three terminals: *base*, *emitter*, and *collector*. When it operates in its linear region, a BJT amplifies its base current, via its internal current gain, to produce a larger collector current of:

$$I_{\text{collector}} = \beta I_{\text{base}}. \qquad (4.87)$$

where β is the current gain.

In a phototransistor irradiation impinges directly at the p–n junctions. One of the junctions is reversed biased, which generates a large current from the changes of the forward biased p–n junction current. The structure of phototransistors is optimized for photo applications and a phototransistor has generally a much larger base and collector areas than that of a normal transistor. However, due to the presence of excessive internal charges the response times of phototransistors are much longer than photodiodes (almost by the β factor).

Old technology of silicon- or gallium-based phototransistors had a *homojunction* structure as can be seen in Fig. 4.52b. However in recent years, in order to ensure higher photon-to-electron conversion and hence larger sensitivity, the emitter contact is often offset within the phototransistor structure as can be seen in Fig. 4.52c. This ensures that the maximum amount of light reaches the active region within the phototransistor.

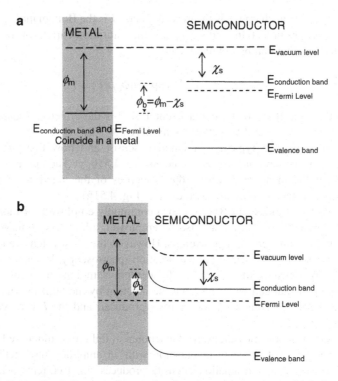

Fig. 4.53 Energy band diagram of a metal and a n-type semiconductor (**a**) before forming a Schottky contact and (**b**) after forming the Scottky contact

4.9.2 Schottky Diode Based Transducers

A diode constructed from a metal–semiconductor junction is called Schottky diode, which is named after Walter H. Schottky, who was a German scientist. Such a metal–semiconductor junction forms a rectifying barrier that only allows current to flow in one direction.

As shown in Fig. 4.53, for a Schottky diode made of n-type semiconductor, the *barrier height*, ϕ_b, depends on the difference between the *work function* of the metal, ϕ_m, and *electron affinity* of the semiconductor, χ_s. Work function of a metal and electron affinity of a semiconductor are associated with the energies, which an electron is required to escape this bands into the *vacuum level*. Before bringing the metal and semiconductor materials in an intimate contact, their *Femi levels* are different (Fermi level is an energy level that an electron with a 50 % chance is above it). After establishing the contact, the Fermi level of the semiconductor forces to become equal to that of the metal, as the system reaches equilibrium after the exchange of free carriers. Consequently, band bending occurs and a barrier height is formed to stop the free carrier exchanges.

Fig. 4.54 A typical response of a Schottky diode based device when current is held constant

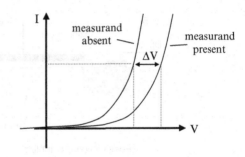

Depending on the selection of the materials utilized in the fabrication of the diode, a change in pressure, ambient temperature, or presence of different gases (which changes the depletion region configuration) causes the *I–V* characteristic of the Schottky diode to change, as seen in Fig. 4.54. The response can be obtained, when the device operates at a constant current and thereby measuring the voltage shift or by operating it in a constant voltage and measuring the current change. Also, by correlating the measured *I–V* characteristics to (4.86), the change in the barrier height in the presence of the measurand can also be experimentally derived.

The semiconducting materials commonly employed for Schottky diodes include silicon, gallium arsenide, and silicon carbide, whilst metals utilized as the Schottky contact include Pd, Pt, and Ni which all have large work functions to establish large barrier heights.

Quite often in chemical sensing applications, a very thin layer, typically a few to tens of nanometers, is added between the metal–semiconductor junction to increase the device sensitivity and selectivity. Semiconducting metal oxide layers, such as SnO_2, TiO_2, WO_3, are popular choices for establishing such layers, as they show high sensitivity towards gases like CO, CH_4, H_2, and O_2.

4.9.3 Field Effect Transistor Based Transducers

A *field effect transistor* (*FET*) is a transducer in which the flow of current, between two of its terminals, is controlled by an applied voltage at a third terminal. As a transducer, it can be employed for converting chemical, physical, and electromagnetic signals into a measurable current.

The most well-known FET is the *metal oxide field effect transistor* (*MOSFET*) shown in Fig. 4.55a. The current flowing between the *drain* and *source* semiconducting electrodes is controlled by the electric field generated by a third electrode, the *gate*, which is located between them. The gate electrode is insulated against drain and source by an oxide layer. The device shown is an *n-channel enhancement MOSFET*. If the gate is made sufficiently positive, with respect to the source (larger than the transistor's *threshold voltage* − $V_{\text{threshold}}$) , electrons are attracted

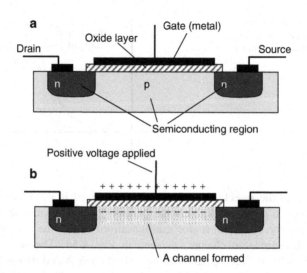

Fig. 4.55 (a) Schematic representation of an n-channel enhancement MOSFET and (b) formation of a conducive layer between drain and source, upon applying a voltage to the gate, that allows the passage of current

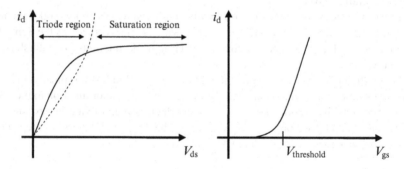

Fig. 4.56 Input and output characteristics of an n-channel enhancement MOSFET

to the region below the gate and a channel of n-type material is created (Fig. 4.55b). This may trigger by the application of a voltage, an external electric field, the adsorption of ions on the gate surface, etc.

The drain–source current, I_{ds}, characteristics of an n-channel enhancement MOSFET (enhanced current by the extra doping of the beneath channel) for different gate–source and drain–source voltages (V_{gs} and V_{ds}) is shown in Fig. 4.56. This current takes on different expressions for two main operating regions and is defined as:

$$I_{ds} = \frac{1}{2}C_{ox}\mu\frac{W}{L}(V_{gs} - V_t)^2 \quad \text{(saturation region)}, \tag{4.88}$$

$$I_{ds} = C_{ox}\mu\frac{W}{L}\left[(V_{gs} - V_t)V_{ds} - \frac{1}{2}V_{ds}^2\right] \quad \text{(triode region)}, \tag{4.89}$$

Fig. 4.57 A schematic
representation of an ISFET

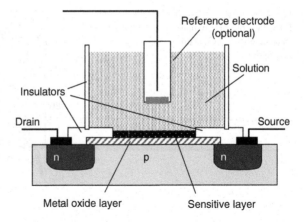

where with C_{ox} is the capacitance of the oxide layer per unit area, W and L are the width and the length of the channel, respectively, and μ is the electron mobility in the channel. V_{gs} and V_{ds} are the current applied between gate and source and drain and source, respectively. The threshold voltage, V_t, is obtained using [12, 13]:

$$V_t = \frac{\phi_m - \phi_s}{q} - \frac{Q_{ox} + Q_{ss} + Q_b}{C_{ox}} + 2\phi_f, \qquad (4.90)$$

where ϕ_m and ϕ_s are the metal and semiconductor work function and electron affinity, respectively. Q_{ox}, Q_{ss}, and Q_b are the accumulated charges in the oxide, oxide/semiconductor interface, and the depletion charge in the semiconductor, respectively. Finally, the last term, ϕ_f, depends on the doping level of the semiconducting material. As a consequence of V_t being dependent on charge and capacitance within the transistor, a FETs I–V characteristics are strongly affected by environmental change in which the device is placed in. Changes in humidity, ions, chemical species, and the presence of dielectric materials in the environment all can alter the characteristics of the device and hence FET can be used for measuring these parameters.

FETs may be implemented for the sensing applications in liquid media, as shown in Fig. 4.57, in which the metal gate is replaced with a reference electrode in a liquid touching the gate metal oxide layer. Insulators are added to protect the connections to the drain and source from the liquid. A sensitive layer is deposited over the oxide layer, forming the sensitive gate. As a result of chemical reactions occurring on this surface, the dimension of the channel between drain and source changes, which results in an alteration of the I_{ds}. The change of current is proportional to the target analyte in the liquid.

The first FET sensors were developed in early 1970 at the Stanford University [12]. They used the device for the measurement of pH. For measuring the redox interactions on the sensitive layer, a reference electrode is needed. An applied voltage with reference to the drain voltage guarantees the formation of an ionic

Fig. 4.58 I_d–V_{ds} curves of an ISFET vs. the pH change (reprinted from [13] with permission)

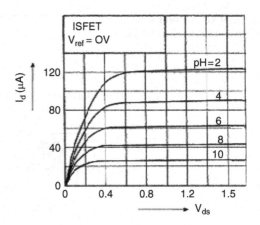

double layer on the surface of gate, which affects the drain–source current. A MOSFET incorporating a reference electrode is used for the measurement of ion concentrations and is called an *ion-sensitive field effect transistor (ISFET)*.

For ISFET the equation of the threshold voltage changes to [12–14]:

$$V_t = E_{ref} - \psi + \chi^{sol} + \frac{\phi_s}{q} - \frac{Q_{ox} + Q_{ss} + Q_b}{C_{ox}} + 2\phi_f. \qquad (4.91)$$

In comparison to (4.90), there are two extra parameters in this equation: the interfacial potential, E_{ref}, and the potential at the solution/oxide interface, and $\psi + \chi^{sol}$. The term $\psi + \chi^{sol}$ consists of ψ, which is a function of the analyte's pH and χ^{sol} is the surface dipole potential. ϕ_m, the work function of the gate metal, does not appear in this equation anymore and its effect is included in E_{ref}.

An example of I_d–V_{ds} curves of an ISFET as a function of the pH of the solution is shown in Fig. 4.58 [13]. The sensor response is due to ψ being a function of pH, which is the chemical input parameter presented as ψ (pH).

An empirical equation for ψ (pH) can be derived from site-dissociation and double-layer models. This equation, which can be described as a sub-Nernstian formula, is a function of pH of the electrolyte. After the simplification, the equation can be presented as:

$$\psi(pH) \approx (-5,916 \text{ mV}) \times (pH_s), \qquad (4.92)$$

in which pH_s is the value of the pH in the vicinity of the sensitive layer surface. When the ion concentration at the surface $[H^+]_s$ changes, the discussion on the effect of diffusion current can be similarly implemented in this case. To extend the dynamic range of ISFETs and/or increase the selectivity, membranes can be added to the gate surface.

Many biosensors based on the chemical reactions on the surface of ISFETs are enzyme based (they will be described in Chap. 5), which are called *enzyme field effect transistors* (*ENFET*). Considerable effort has been made to develop different types of ENFETs since the introduction of an ENFET for sensing penicillin [15]. The performance of ENFET biosensors is greatly affected by the integration mechanism of the enzymes with the ISFET. Immobilizing enzymes on the gate surface of ISFETs is a routine method to establish a sensitive layer.

4.10 Acoustic Wave Transducers

Acoustic wave transducers are constructed based on piezoelectric materials within which mechanical waves are launched. As described in Chap. 3, piezoelectric phenomenon occurs in crystals that do not have a center of symmetry. Applying stress to such crystals deforms their lattices and electric fields are produced, and vice versa. It is believed that the first acoustic wave device was developed in early 1920s by Walter Cady who was an American engineer. It was a resonator element made of *quartz*, which was used for stabilizing electronic oscillators.

Generally propagation of acoustic waves in solids can be divided into two categories: *bulk acoustic waves* (*BAWs*), in which acoustic waves propagate in the bulk of a solid, and *surface acoustic waves* (*SAWs*) , in which their propagation is confined within a region near the surface of the solid.

As the acoustic wave propagates through the bulk or on the surface of the material, any changes to the characteristics of the propagation path affect the velocity (hence phase) and amplitude of the waves. These changes (perturbations) can be monitored by measuring the frequency or phase characteristics. When employed for sensing, such perturbations are measured. Consequently, they can be correlated to the corresponding physical or chemical quantities being measured.

4.10.1 Quartz Crystal Microbalance

The *quartz crystal microbalance* (*QCM*) is the most commonly BAW resonator that is based on quartz crystal, which was one of the first piezoelectric material that was discovered and incorporated into devices. The schematic representation of a typical QCM is shown in Fig. 4.59. Generally, a quartz substrate is cut into thin disks with metal pads deposited on both sides, at which electrical signals are applied.

The piezoelectric crystal transforms the applied electric signal on the metal pads to acoustic waves. These acoustic waves are bounced within the top and bottom boundaries, reflecting back and forth in the bulk of the quartz crystal, resulting in resonation. The addition of mass on the surface of the quartz crystal increases its thickness. As a result, the resonator holds longer wavelength standing waves, which accommodates smaller resonant frequencies (Fig. 4.60).

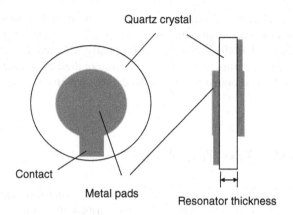

Fig. 4.59 Top and side views of a QCM

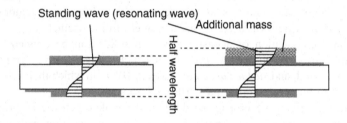

Fig. 4.60 The effect of resonator thickness increase on the oscillation cavity

The relationship between the change in the oscillation frequency, Δf, of a QCM to the change in mass added to the surface of the crystal, Δm, is given by the *Sauerbrey equation*, which was first presented by G. Sauerbrey in 1959:

$$\Delta f = \frac{-2\Delta m f_0^2}{A\sqrt{\rho\mu}} = \frac{-2\Delta m f_0^2}{A\rho v},$$ (4.93)

where f_0 is the resonant frequency of the crystal, A is the area of the crystal, and ρ, μ, and v are the density, shear modulus, and shear wave velocity of the substrate, respectively. As can be seen, any increase in Δm results in a decrease in operational frequency Δf.

Clearly the oscillating frequency's dependence on mass change makes the QCM ideally suited for sensing applications. The mass sensitivity can be defined as the change in frequency per change in mass on the unit area of the device. QCM mass sensitivity can be enhanced by adding a sensitive layer on its surface. As observed in (4.93), increasing the operational frequency (or the reduction in the crystal thickness) will increase the QCM's sensitivity.

Example 2. Consider a QCM with an operational frequency of 5 MHz. If the shear modulus is 3×10^{11} g cm^{-1} s^2, the density of quartz is 2.6 g cm^{-3}, what is the frequency change upon the added mass of 1 ng? The surface area of your device is 0.25 cm^2.

Answer: According to the Sauerbrey equation:

$$\Delta f = \frac{-2 \times [10^{-9} \text{ (g)}] \times [5 \times 10^6 \text{ (Hz)}]^2}{[0.25 \text{ (cm}^2)] \times \sqrt{[2.6 \text{ (g cm}^3)] \times [3 \times 10^{11} \text{ (g/cm}^{-1} \text{s}^2)]}} = 0.225 \text{ (Hz)}.$$

The *Q factor* of a QCM, which is the ratio of frequency and bandwidth, can be as high as 10^6. Such a large Q leads to a highly stable oscillation, for which the resonance frequency can be accurately determined. Commonly, a resolution as small as 0.1 Hz for a crystal operating at a fundamental resonant frequency of approximately 5 MHz can be obtained. Considering such frequency stability, for a 5 MHz device with a surface area of 0.25 cm^2 operating in air, the mass detection limit of a QCM is less than 2.2 ng cm^{-2}.

In addition to sensing mass changes, the electrical boundary condition perturbations also change the piezoelectric properties of a device and therefore its resonant frequency. Consequently, such a device can be used for monitoring charge and conductivity changes as well as mass changes.

QCMs have been employed for measuring mass binding from gas-phase species for moisture and volatile organic compounds sensing and environmental pollutants. They have also been employed for monitoring gas concentrations, redox reactions occurring on metal oxide or polymeric sensitive layers deposited on top of them. QCMs operating in liquid media were adapted in 1980s to measure viscosity and density of liquid in contact with them. They have also been successfully realized as commercially available for biosensing applications. Examples concerning such biosensors will be presented in Chap. 5.

The efficiency with which a piezoelectric material can transform mechanical waves into electromagnetic waves, and vice versa, is assessed by its piezoelectric coupling coefficient, k^2. In the past few decades, several crystals have emerged that exhibit larger k^2 than quartz, These include: *lithium niobate (LiNbO₃)*, *lithium tantalite (LiTaO₃)*, and more recently *lanagsite (LNG)*. LiNbO₃ and LiTaO₃ have k^2s that are almost an order of magnitude larger than that of quartz. These materials, however, do have drawbacks, such as not being suitable for fabricating bulk type devices due to their fragile structure and low Q. However, they are being rigorously investigated for SAW applications.

4.10.2   Film Bulk Acoustic Wave Resonator

Film bulk acoustic wave resonators (FBARs) represent the next generation of bulk acoustic wave resonators after the QCMs. They are composed of thin films and have

Fig. 4.61 Schematic of the cross-sectional view of a typical FBAR used as a sensor

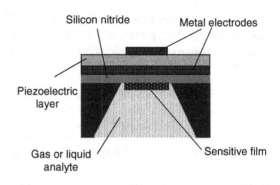

much smaller dimensions than QCMs. They have a relatively high operating frequency, which is the main reason for their higher sensitivity. In addition, their fabrication is compatible with current standard *microelectromechanical system* (*MEMS*) technologies. A typical FBAR is built on a low-stress and inert membrane (such as silicon nitride) together with a piezoelectric film, which is sandwiched between two metallic layers (Fig. 4.61). Silicon nitride is commonly used in the fabrication of FBARs as it facilitates the etching process. While silicon substrate is etched in strong alkaline etchants such as potassium hydroxide (KOH), it does not etch *silicon nitride*.

FBARs generally have a resonant quality factor Q in a range of 20–1,000 at operational frequencies of 1 GHz. This Q is much lower than their QCM counterpart. Unfortunately, this low Q translates into a noisier and less stable oscillation. This reduces the overall detection limit of the system.

If the membrane thickness is small with respect to the acoustic wavelength, then the frequency change Δf, resulting from added mass, Δm, of adsorbed target analyte, can be found from the *Lostis approximation* according to:

$$\frac{\Delta f}{f_0} = -\frac{\Delta m}{m_0},\qquad(4.94)$$

where $m_0 = \rho S d$ is the mass of the resonator in which ρ is the density, S is the surface area, and d is the thickness of the membrane, and $f_0 = v_p/2d$, with v_p being the acoustic wave phase velocity. The fractional frequency change $\Delta f/f_0$, produced by the mass of analyte adsorbed per unit surface ($\Delta m/S$), is then equal to:

$$\frac{\Delta f}{f_0} = -\frac{1}{\rho d}\left(\frac{\Delta m}{S}\right).\qquad(4.95)$$

As a result, the frequency shift is:

$$\Delta f = -\frac{v_p}{2\rho_p d^2}\left(\frac{\Delta m}{S}\right).\qquad(4.96)$$

Table 4.2 Acoustic phase velocity and density for three different piezoelectric materials: Frequency shifts produced by 1 ng cm^{-2} of adsorbed analyte are calculated for a thickness d

Material	v_p (m s^{-1})	ρ_p (kg m^{-3})	(Hz)
Quartz ($d = 100$ μm)	3,750	2,648	0.708
AlN ($d = 1$ μm)	11,345	3,260	34,800
ZnO ($d = 1$ μm)	6,370	5,665	11,244

From (4.96), for a given $\Delta m/S$, the frequency shift magnitude increases with increasing the acoustic wave velocity in the piezoelectric resonator medium and decreasing its density. Additionally, the frequency shift magnitude increases quadratically as the membrane thickness decreases.

Example 3. Use the density and phase velocity of the FBARs of different thicknesses presented in Table 4.2 to calculate the frequency shift produced by the adhesion of 1 ng cm^{-2} of analyte on the surface of the sensor for the three different piezoelectric materials (quartz, AlN, and ZnO) presented in the table for their corresponding thicknesses.

Answer: The frequency shifts due to the addition of 1 ng cm^{-2} of analyte are presented in the table. As can be seen, the materials composed of bulk (AT-quartz), thin film (AlN, ZnO) resonators, with the thickness of $d = 100$ and 1 μm, respectively. The ZnO and AlN thin film resonators show frequency shifts of approximately 16,000 and 5,000 Hz, respectively, in comparison with AT-quartz resonator shows a frequency shift of only 0.708 Hz.

The above example obviously shows the advantage of using FBARs as transduction platforms over conventional QCMs (such as the one presented as the bulk resonator based on quartz) in terms of much larger frequency shift for the same added mass. However, it should be considered that as the stability of the FBARs is much less than that of the QCMs, the mass detection limits of FBARs are still not far better than those of QCMs.

4.10.3 Cantilever-Based Transducers

Transducers consist of microcantilevers are designed to resonate at certain frequencies or produce measurable deflections upon exposure to different target stimuli (Fig. 4.62). The behavior of these resonant frequencies and deflections is governed by the incorporated materials in their structures and their dimensions. The resonant frequency and deflection changes can be due to environmental stimuli such as the adhesion of particles on the cantilevers' surfaces and changes in the physical properties of the environment (i.e., temperature, stress, electric field, and viscosity variations). Cantilevers' resonant frequencies are typically in the range of 100 MHz to 5 GHz.

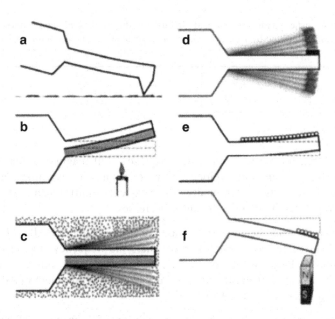

Fig. 4.62 Possible uses of a cantilever transducer (*side view*) for the measurement of different properties: (**a**) force; (**b**) temperature, heat; (**c**) medium viscoelasticity; (**d**) mass (end load); (**e**) applied stress, and (**f**) magnetic measurements of magnetic beads on its surface (reprinted from [16] with permission)

Cantilevers are great sensors for low quantities of masses. In their oscillation mode, they can operate in certain resonant modes and the attached masses onto their surface can alter the resonant frequencies by shifting them. Using the conventional harmonic oscillator formula, the resonant frequency, f, of a cantilever is relevant to its spring constant, K, as:

$$\omega = 2\pi f = \sqrt{\frac{K}{m}}. \tag{4.97}$$

in which m is its mass. K of a cantilever is related to its dimensions via:

$$K = \frac{EWt^3}{4L^3}, \tag{4.98}$$

in which W is the width of the cantilever, E is the *Young's modulus* (Young's modulus is a measure of the stiffness of an elastic material), L is the cantilever length, and t is its thickness.

Using (4.97) the shift in resonant frequency as a function of the added mass can be calculated as:

$$\Delta m = \frac{K}{4\pi^2} \left(\frac{1}{f_1^2} - \frac{1}{f_0^2} \right), \tag{4.99}$$

where f_0 and f_1 are the operating frequencies before and after adding the mass, respectively.

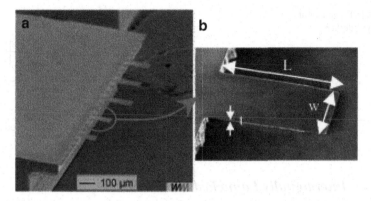

Fig. 4.63 SEM images: (**a**) five silicon-based cantilevers, (**b**) zoomed image of a cantilever with length L of 100 µm, width W of 40 µm, and thickness t of 0.5 µm. Reprinted with permission from Elsevier publications (reprinted from [16] with permission)

Many cantilevers operate based on bending modes. In this case, the thin bar is used to measure changes in the differential surface stress of its opposite sides. These days, such cantilevers are commonly applied in atomic force microscopy measurements. The *Stoney's formula* can be used for calculating the amount of deflection Δz at one end of the cantilever, when a change in stress $\Delta \sigma$ is applied:

$$\Delta z = \frac{3(1 - \nu)L^2}{Et^2}(\Delta \sigma), \tag{4.100}$$

where ν is *Poisson's ratio* (Poisson's ratio is a measure of the Poisson effect. This effect describes that when a material is compressed in one direction, how it expands in the other directions).

Cantilevers can be implemented in a variety of sensing measurements as demonstrated in Fig. 4.62. The cantilever may be made from layers having different thermal expansion coefficients, in which case a temperature change causes it to bend. Such cantilevers are capable of measuring temperature changes as small as 10^{-5} K. A shift in resonance frequency may also arise from changes in medium viscosity or mass added to the cantilever's surface. Increasing the viscosity, as well as adding a mass, dampens a cantilever's oscillation, lowering its operational frequency. As mass sensors, microcantilevers can be employed for monitoring biological interactions such as: enzymatic and antigen–antibody interactions as well as interactions of complementary DNA strands which is described in Chap. 5. SEM images of microfabricated silicon-based rectangular cantilevers of different lengths are shown in Fig. 4.63.

The minimum detectable mass of a cantilever can be as low as $\approx 10^{-15}$ g, which is excellent in comparison to other acoustic wave devices. Unfortunately, when operating in liquid media, a cantilever's resonant frequency shifts toward lower values and the quality factor (Q) decreases dramatically, as a result of damping caused by the liquid. This means significant loss of detection limit for the cantilever operating in liquid media.

Fig. 4.64 Basic layout
of a SAW device

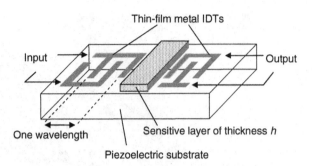

Thin-film metal IDTs

Input

Output

One wavelength

Sensitive layer of thickness *h*

Piezoelectric substrate

4.10.4 Interdigitally Launched SAW Devices

Both *SAWs* and *BAWs* can be launched using *inter-digital transducers* (*IDTs*), which were first incorporated for such a task by White and Voltmer from Berkeley University, USA, in early 1960s. In such devices, metal thin film IDTs are patterned on the surface of a piezoelectric crystal substrate of a special cut that allows the propagation of either SAWs or BAWs or both. As shown in Fig. 4.64, when an oscillating voltage is applied to the input IDTs, they launch acoustic waves. They travel along the surface or within the bulk (in this case, they have to bounce back from the bottom surface of the substrate) to the output IDTs, where the acoustic waves are converted back into electric signals. A sensitive layer may be deposited on the active area of the device to provide sensitivity to target analytes.

Depending on the crystal cut and IDTs geometry, different propagation modes are possible. The wave can propagate via transverse or longitudinal modes or a combination of both. For instance, for *Rayleigh waves*, the particles near the surface move elliptically along the propagation direction, normal to the surface and for *shear horizontal waves* (*SH*), particles near the surface are displaced parallel to the surface.

The development and utilization of different piezoelectric crystals and investigation of wave propagation within different crystal orientations have resulted in the discovery of various acoustic wave propagation modes. These include: *leaky SAW* (*LSAW*) and *surface-skimming bulk wave* (*SSBW*), which both are shear horizontal SAWs. Other major acoustic propagation modes that can be launched and received using IDTs include: *acoustic plate mode* (*APM*) and *Lamb wave* (or *flexural plate wave—FPW*), which are both BAWs.

In addition to different crystal types and orientations, the modes of propagation can be altered by the deposition of layers onto piezoelectric substrates. For instance, if the propagation velocity of the shear horizontal (SH) waves in the deposited layer is less than that of the substrate, the SH wave in the substrate can transform into a near surface confined wave (Fig. 4.65). This horizontally polarized shear mode of operation is generally referred to as Love mode (named after Augustus Love, who theoretically suggested their existence in 1911). The Love mode SAW sensors are the most sensitive devices in the acoustic wave family. Sensitivity of a Love mode

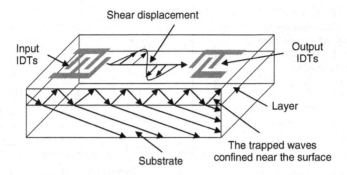

Fig. 4.65 Basic layout of a Love mode SAW device

SAW sensor, based on 90° rotated ST-cut quartz crystal operating at 100 MHz, can be up to two orders of magnitude greater than a QCM operating at 10 MHz, which is the result of the near the surface confinement of acoustic wave propagation.

SAW devices can be utilized for sensing various physical and chemical parameters including temperature, acceleration, force, pressure, electric fields, magnetic fields, ionic species, gas flow, vapor concentration, viscosity, and in biosensing.

SAW devices can be employed as affinity sensors through the addition of a mass sensitive layer on their active surface area. The following equation describes the frequency shift, Δf, as a function of the added mass on the active area of the device (Fig. 4.64), which is calculated using the perturbation theory [17, 18]:

$$\Delta f \approx S f_0{}^2 h \rho, \tag{4.101}$$

where f_0 is the operational frequency, S is a parameter extracted from the material constants of the substrate, ρ is the added coating layer, and h is the thickness of the added layer. It can be seen from the equation that there is a linear relationship between the thickness and density of the added layer and the frequency shift, which is the base of the device operation as a mass sensor. Obviously the frequency shift is also proportional to $f_0{}^2$, which means that devices operating at higher frequencies show better sensitivities.

SAW devices can also be used for monitoring changes in conductivity and charge of a sensing layer. Many metal oxides and conductive polymers change their conductivity in response to different oxidizing or reducing species. Depositing such materials over the active area of a SAW device can transform it into a device sensitive to such alterations. This property originates from the fact that the piezo-electricity is a phenomenon, which relates acoustic waves to changes in electric field. Again using the perturbation theory, the shift in operational frequency, f, after adding the conductive layer is given by [18]:

$$f = f_0 \frac{k^2}{2} \frac{1}{1 + (\sigma_{SH}/\sigma_{OR})^2}, \tag{4.102}$$

Fig. 4.66 Maximum
response occurs when sheet
conductivity value is matched
to product of SAW mode
velocity and substrate
permittivity

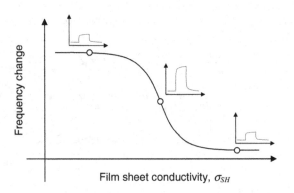

where f_0 is the operational frequency, k^2 is the electromechanical coupling coefficient, σ_{SH} is the sheet conductivity of the sensitive layer, σ_{OR} is the multiplication of SAW mode velocity and the substrate permittivity (which has a similar unit as sheet conductivity). Obviously, a SAW device with a higher electromechanical coupling coefficient generates a larger frequency shift in response to target analytes, hence is more sensitive.

The operation point at which the maximum frequency shift can be calculated from the slope of the graph, which is obtained by plotting f in (4.102) against σ_{SH}, is demonstrated in Fig. 4.65. From the tangent of this curve, maximum sensitivity is extracted when σ_{OR} and σ_{SH} are equal (Fig. 4.66).

For sensing in liquid media, SH-SAW modes are preferred as they suffer much less attenuation when liquid comes in contact with the propagating medium. As the particle displacements are shear horizontal, not normal to the sensing surface, the contact liquid cannot damp the movements, except in cases where the liquid has high viscosity. This makes them ideally suited for biosensing applications. In such devices, a bio-selective layer is deposited on the surface of the device.

IDTs can also be designed for launching and receiving Lamb waves which are BAWs. Lamb waves propagate in plates (diaphragm) and in nature are composed of two Rayleigh waves propagating on each side of the plate. Two groups of Lamb waves can propagate through the plate independently, which include symmetrical and asymmetrical waves (Fig. 4.67).

Due to the low phase velocities required to enable low-loss wave propagation, the operating frequency of Lamb waves generally falls within the range of 5–20 MHz. The mass sensitivity of a Lamb mode sensor is given by:

$$S = \frac{-1}{2\rho d},\tag{4.103}$$

where d is the plate thickness and ρ is the density of the diaphragm. Mass detection limit of a 10 MHz device with the thickness of 2 μm can be as low as 200 pg cm^{-2}.

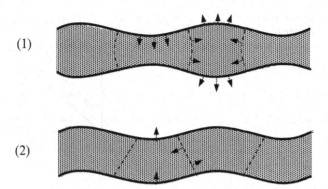

(1)

(2)

Fig. 4.67 (1) Symmetric and (2) asymmetric Lamb waves. *Arrows* depict the direction of the particle displacements during the wave propagation

4.11 Gyroscopes

Gyroscopes are devices that can measure angular velocities and play an important part in navigations and increasing in consumer electronics. As early as the early eightieth century, crude spinning devices were being used for sea navigation. However, more traditional spinning gyroscope was invented in the early 1800s. Later in 1910s, gyroscopes were adapted for navigations in aircrafts. In the progress, optical gyroscopes were invented in 1960s. However in the last two decades, MEMS gyroscopes were introduced and mass-produced successfully for sensing applications in products such as *global positioning systems (GPSs)*, for interactive game consoles and smart phones.

The functionality of a gyroscope depends on its type. Spinning gyroscopes operate based on a spinning object, which is tilted with reference to the direction of the spin. Such an object shows a precession under a secondary force such as gravity. This spinning object, instead of falling, defies gravity and the free end of the axis slowly describes a circle in a horizontal plane. The precession keeps the device oriented and the angle relative to the reference surface can be measured. To understand how a gyroscope operates the reader should first know the meanings of *torque* and *angular momentum*. Torque, τ, is obtained as a result of a force, F (that causes a rotation), using a vector cross product (Fig. 4.68):

$$\tau = r \times F \quad \text{for which} \quad |\tau| = |r||F|\sin\theta, \tag{4.104}$$

where r is the displacement vector (a vector from the point from which torque is measured to the point where force is applied), and θ is the angle between the position and force vectors. Angular momentum, L, is defined as:

$$L = r \times p, \tag{4.105}$$

Fig. 4.68 Illustration of the torque vector definition

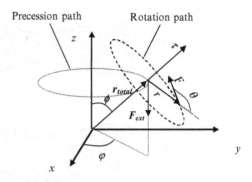

where p is the momentum of the object as $m \cdot V$, for a particle with a mass of m and $F = m \cdot dV/dt$. As a result for a particle circling at a constant distance from a center, torque will be the change in angular momentum in time as:

$$\tau = \frac{dL}{dt}. \tag{4.106}$$

Subsequently, L and τ vectors have both similar directions. L is defined as $L = I\omega$, where I is the *moment of inertia* and ω is the angular velocity. This means that torque is a function of ω as:

$$\tau = I\frac{d\omega}{dt}. \tag{4.107}$$

Imagine that a particle is spinning around a center with a torque of τ. If a second torque τ_{ext} as a result of an external force F_{ext} (as shown in Fig. 4.68) appears, a rotation about the axis perpendicular to both τ_{ext} and L occurs and this generates the precession. In a spin gyroscope the gravity (the gravity force of mg) acts downward on the device's center of mass, and an equal force acting upward to support one end of the device, where it is touching the ground. The rotation resulting from this torque causes the device to rotate slowly about the supporting point. The resulting precession angular frequency, Ω_p, is:

$$\tau_{ext} = \Omega_p \times L. \tag{4.108}$$

Equation (4.107) is satisfied only if:

$$|\Omega_p| = \frac{r_{total}F_{ext}}{I\omega}. \tag{4.109}$$

Obviously, change in F_{ext} can cause the angular velocity of the precession to change. This means by measuring the angular velocity of the precession, the applied external force, or as a result the acceleration of the system, in which the gyroscope is placed within, can be estimated.

A rotating gyroscope can be used as a tilt sensor as it resists any changes against its orientation. Initially, the orientation sensors were made of rotating gyroscopes in

Fig. 4.69 Schematic
of a vibrating gyroscope

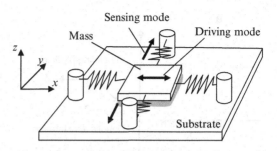

cages that could allow their free rotation. If the main object was tilted, the gyroscope would still be holding its position and the degree of tilt could be measured using a reference sensor.

Optical gyroscopes are based on the interference of two initially coherent laser beams. They send the two beams around a circular path in opposite directions. If the path spins, a phase shift can be detected since the speed of light always remain constant. Such gyroscopes are usually used in airplanes.

MEMS gyroscopes are the most commonly used gyroscopes these days. They are inexpensive vibrating structures manufactured using MEMS technology. They are packaged similar to other integrated circuits and may provide either analog or digital outputs. MEMS gyroscopes use lithographically constructed configurations such as *tuning forks*.

As the fabrication of low-friction bearings in microdimensions is impractical, reducing the size of classical gyroscopes that incorporate a spinning wheel is not possible. In MEMS instead, a mass mounted on a spring suspension is utilized for making micro-gyroscopes. In such structures, the mass vibrates back and forth in translational motion, as schematically demonstrated in Fig. 4.69.

In the very basic configuration depicted in Fig. 4.69, the mass is first make to oscillate in the x-axis, which is called the *drive axis*. Once in motion, the mass is sensitive to angular rotation about the z-axis. The presence of an angular rotation thus induces a *Coriolis acceleration* in the y-axis, which is called the sense axis. The presence of a stress transducer (such as a piezoresistance material) will allow the measurement of this force.

All vibrating gyroscopes operate based on the Coriolis acceleration. This acceleration is experienced by a mass that undergoes a linear motion in a frame of reference that is rotating about an axis perpendicular to that of the motion of the mass. As a result an acceleration, which is directly proportional to the rate of turn, is seen by an observer who is anchored to the axis.

To understand the Coriolis effect, imagine that a mass traveling with a constant velocity of v as presented in Fig. 4.70a. An observer attached to the xyz-axis of the coordinate system is watching this body. If the coordinate system starts rotating around the z-axis with an angular velocity of Ω, to the observer the body is changing its trajectory toward the z-axis with an acceleration of $2v \times \Omega$. Although no real force has been exerted on the body, to the observer, an apparent force has resulted that is directly proportional to the rate of rotation. This effect is the basic operating principle of vibratory structure gyroscopes.

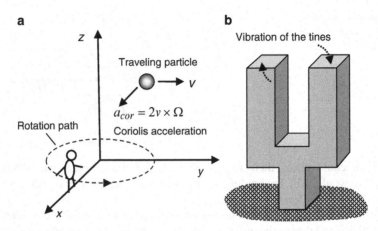

Fig. 4.70 (a) The Coriolis effect. (b) Tuning-fork vibratory gyroscope. The tines are differentially driven to a fixed amplitude. Coriolis force is detected either as differential bending of the tuning-fork tines or as a torsional vibration of the tuning-fork stem

One of the most widely used MEMS gyroscopes is the tuning fork design from the Charles Stark Draper Laboratory in Cambridge Massachusetts in early 1990s (schematic presented in Fig. 4.70b). The design consists of prongs connected to a junction bar (only two prongs are demonstrated here), which resonate at certain amplitude. When the prongs rotate, Coriolis force causes a force perpendicular to the prongs of the fork. The force is then detected as bending of the tuning fork. These forces are proportional to the applied angular rate, from which the displacements can be measured. A scanning electron micrograph of an actual gyroscope for measuring torsional acceleration is shown in Fig. 4.71, which is fabricated using MEMS techniques onto a silicon wafer.

The first MEMS type gyroscopes were developed for the automobile industry. These devices were implemented as the angular velocity sensors for skid control in anti-lock braking applications in cars. Nowadays they are used in all different kinds of home electronic appliances and smart phones.

4.12 Summary

Some of the most popular transduction platforms were introduced in this chapter. The chapter started with the description of interdigitated electrodes, between which a sensitive layer is placed. It was shown that the addition of a sensitive layer makes them useful for conductometric and capacitive sensing measurements.

Optical waveguide based transducers were then presented, and it was shown that such transducers have various configurations. In sensing applications, a sensitive layer is generally deposited over the waveguide layer. As was shown, the sensitivity is a function of the effective refractive index of the layer/sample medium that is in contact with the waveguide. Optical transducers are widely used in chemical

Fig. 4.71 A capacitive torsional accelerometer (reprinted from [19] with permission)

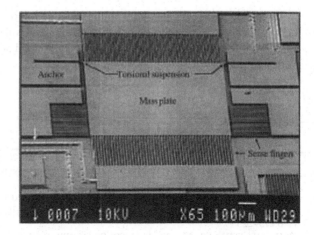

sensing for medical and biochemical applications. SPR technology has become quite mature nowadays, and there are many commercially available SPR systems.

The sensor platforms based on spectroscopy systems were introduced, including vibrational spectroscopy platforms which incorporate infrared and Raman systems. The spectroscopy systems, in which light affects the materials according to the bandgap, including UV–Visible spectroscopy and PL spectroscopy units were presented and eventually the applications of NMR units were briefly illustrated.

Electrochemical transducers were described next. They are the most commonly employed sensors in biosensing, and additionally are suitable for a wide range of applications, including ion concentration measurements; sensing environmental pollutants, and gas concentrations. Such transducers not only offer a myriad of sensing applications but also provide a means for investigating the many physiochemical interactions that occur during chemical interactions, in particular, those occurring at the surfaces of the electrodes.

Solid state transducers were presented. They are versatile and consist of devices based on junctions between metals, semiconductors, and insulators. They can also be found in numerous configurations, such as BJT, MOSFET, and ISFET. Changes in the ambient, both physical and chemical, result in changes in the junction depletion region or the distribution of field. Semiconductor–semiconductor junction diodes and transistors are extensively used as optical sensors, while metal–semiconductor junction devices are popular for gas sensing applications. Devices exploiting field effects are becoming popular for chemical and biochemical applications as practical drawbacks of the technology are being addressed.

Acoustic wave transducers were introduced next. They are very useful as mass microsensors. Different configurations, including SAW, BAW, and cantilever type acoustic wave transducers are particularly useful for sensing application in liquid media. However, they are also incorporated in conductometric, thermal, and mechanical measurements. Love mode devices and thin film bulk resonators are becoming popular as their fabrication costs decrease, offering much higher sensitivities than QCMs.

The chapter was wrapped up with the introduction of gyroscope sensors, which are very popular in vehicles and household appliances these days. Their operation was described and their applications in navigation and acceleration measurements were signified.

References

1. Curkov LM, McCromick PG, Galatsis K, Wlodarski W (2001) Gas sensing properties of nanosized tin oxide synthesised by mechanochemical processing. Sensors Actuators B Chem 77:491–495
2. Lukosz W (1991) Principles and Sensitivities of Integrated Optical and Surface- Plasmon Sensors For Direct Affinity Sensing and Immunosensing. Biosensors Bioelectronics 6:215–225
3. Homola J, Yee SS, Gauglitz G (1999) Surface Plasmon resonance sensors: review. Sensors Actuator B Chem 54:3–15
4. Kretschmann E (1971) Die Bestimmung optischer Konstanten von Metallen durch Anregung von Oberfldchenplasmaschwingungen. Physik Z 241:313–324
5. Yaacob MH, Breedon M, Kalantar-zadeh K, Wlodarski W (2009) Absorption spectral response of nanotextured WO_3 thin films with Pt catalyst towards H_2. Sensors Actuators B Chem 137:115–120
6. Prodi L, Bolletta F, Montalti M, Zaccheroni N (1998) A Fluorescent sensor for magnesium ions. Tetrahedron Lett 39:5451–5454
7. Tao A, Kim F, Hess C, Goldberger J, He R, Sun Y, Xia Y, Yang P (2003) Langmuir-Blodgett silver nanowire monolayers for molecular sensing using surface-enhanced Raman spectroscopy. Nano Lett 3:1229–1233
8. Skoog DA, Holler FJ, Crouch SR (2006) Principles of instrumental analysis, 6th edn. Brooks Cole, Belmont
9. Wang J, Bhada RK, Lu J, MacDonald D (1998) Remote electrochemical sensor for monitoring TNT in natural waters. Analytica Chimica Acta 361:85–91
10. Li J, Peng T, Peng Y (2003) A Cholesterol Biosensor Based on Entrapment of Cholesterol Oxidase in a Silicic Sol-Gel Matrix at a Prussian Blue Modified Electrode. Electroanalysis 15:1031–1037
11. Sze SM, Ng KK (2006) Physics of semiconductor devices, 3rd edn. New York, Wiley-Interscience
12. Wise KD, Angell JB, Starr A (1970) An integrated-circuit approach to extracellular microelectrodes. IEEE Trans Biomed Eng 17:238–247
13. Bergveld P (2003) Thirty years of ISFETOLOGY: What happened in the past 30 years and what may happen in the next 30 years. Sensors Actuators B Chem 88:1–20.
14. Bergveld P (1970) Development of an Ion-Sensitive Solid-State Device for Neuro-Physiological Measurements. IEEE Trans Biomed Eng 17:70–71
15. Caras S, Janata J (1980) Field-effect transistor sensitive to penicillin. Anal Chem 52:1935–1937
16. Raiteri R, Grattarola M, Butt HJ, Skladal P (2001) Micromechanical cantilever-based biosensors. Sensors Actuators B Chem 79:115–126
17. Ballntine DS, Wohltjen H (1989) Surface acoustic wave devices for chemical analysis. Anal Chem 61:A704–A706
18. Ricco AJ, Martin SJ, Zipperian TE (1988) Surface acoustic wave gas sensors based on film conductivity changes. Sensors Actuators B Chem 8:978–984
19. Yazdi N, Ayazi F, Najafi K (1998) Micromachined inertial sensors. Proc IEEE 86:1640–1659

Chapter 5
Organic Sensors

Abstract The importance of organic materials, with an emphasis on biomaterials, in sensing applications will be illustrated. The devices presented in this chapter are either directly used for sensing organic materials or have organic materials in their structures. Different methods of surface modification techniques, which allow and enhance the interaction of such organic materials with the active areas of transducers, will be presented. The most popular organic sensors will be described next.

5.1 Introduction

Some of the most important applications of sensors emerge in *bio-component sensing*. As all bio-components, more or less incorporate organic elements, sensors dealing with organic materials encompass this family. Organic materials are those which are composed of organic compounds. To date, still there is no official definition for organic compounds. Some textbooks define them as materials containing one or more C–H bonds; others include C–C bonds in the definition.

Organic sensors exploit organic materials, in particular *biomaterials*, as components of transducers and/or their sensitive layers. Organic molecules of interest in the fabrication of such sensors consist of biomolecules such as *nucleic acids* (DNA, RNA, etc.), *proteins* (antibodies, enzymes, etc.), other *biomolecules* (carbohydrates, lipids, peptides, metabolites, etc.), and a myriad of *natural* and *synthetic organic materials* of various sizes and lengths (hydrocarbons, polymers, dendrimers, etc.).

This chapter will focus on organic sensors. We will start with a brief look at the properties of different surfaces used in organic sensing applications and the molecular interaction of organic materials with such surfaces. The properties of the most commonly utilized surfaces, such as gold, silicon, glass, as well as conductive and

K. Kalantar-zadeh, *Sensors: An Introductory Course*,
DOI 10.1007/978-1-4614-5052-8_5, © Springer Science+Business Media New York 2013

nonconductive polymers will be explained and the readers will learn how to incorporate them into sensors. Next, some of the most popular organic macromolecules will be introduced for their importance in the development of sensors and sensitive layers. In these sections, proteins, DNAs, and RNAs, as the most important bio-components for sensing applications, will be briefly described.

Throughout this chapter, the importance of biomolecules for organic sensing will be signified. The use of organic and assembly of organic/inorganic materials as the basis of organic sensors will be presented.

5.2   Surface Interactions

If the sensor is a *surface type* device, its active area is prepared by the incorporation of sensitive and selective layers onto the surface. In an organic sensor, these layers are either made of organic materials or the layers are inorganic but have the ability to target organic materials. It is similar in the case of *bulk type* organic sensors. In this case, either the bulk of the sensor has organic components or the target analyte is organic. Interestingly, the bulk type sensors also deal with surfaces of grains or molecules within bulk of the transducer or media in order to operate. Therefore, comprehensive knowledge about the surface of molecules and grains is equally important in the design and implementation of bulk type sensors.

In this section, we will look at the most common processes for the creation of surfaces suitable for the interaction of organic, and in particular, biomolecules.

5.2.1   Targeting and Anchoring Organic Molecules

The literature on the targeting of organic molecules, in particular, biomolecules and anchoring them onto a surface is quite extensive. It is important to remember that in the process many issues relating to maintaining activity, optimizing accessibility, and materials should be addressed [1].

Different approaches are possible for the development of sensitive surfaces and manipulating them. The approach depends on the transducers' types, the nature of organic components, surface chemistry of the transducers, the way devices are used, and the nature of the target molecules. The process of targeting and anchoring organic molecules onto transducers' surfaces can be conducted using many different techniques including:

Adsorption: Chemical components can be adsorbed onto different surfaces by strong forces such as *ionic* (electrostatic forces when opposite charges attract each other) and weaker forces such as *Van der Waals* [force between: two permanent dipoles (*Keesom force*); a permanent dipole and a corresponding induced dipole (*Debye force*); and two instantaneously induced dipoles (*London dispersion force*)] (Fig. 5.1). Most of the proteins can be directly adsorbed onto the surface of

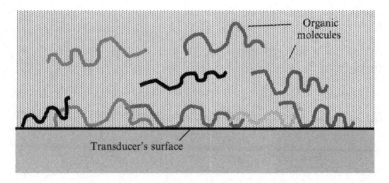

Fig. 5.1 Schematic depicting the surface adsorption of organic molecules via direct adsorption

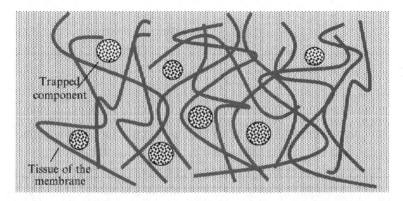

Fig. 5.2 Schematic depicting the physical entrapment of chemical components within the matrix of a membrane

carbon or gold. Many other organic molecules also show similar tendency. Adsorption is inexpensive and simple process to implement for many transducers. However, it can be relatively unstable, not selective, may cause the protein denaturization (when a protein loses its fundamental structure), and its formation depends on the type of analytes used.

Physical entrapment: In this method, generally a semipermeable membrane is used for trapping chemical components. Small components can diffuse freely in and out of bulk of a membrane, but larger molecular weight components are trapped (Fig. 5.2). Entrapment materials can be used in mass production and have great applications in separation and filtering of organic components. However, it is generally complicated to engineer highly selective membranes and the diffusion of particles into the bulk of the membrane can be slow, which might have a deteriorating effect on the sensors' responses incorporating them.

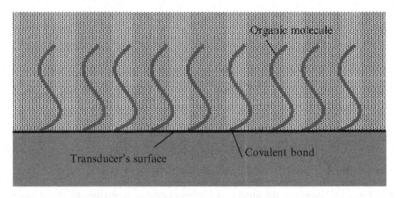

Fig. 5.3 Covalent coupling onto the surface

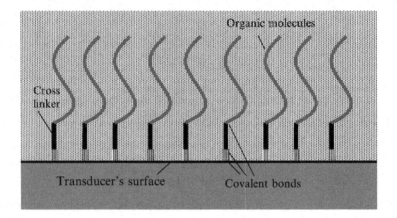

Fig. 5.4 Cross linking using two organic molecules and a cross linker

Covalent coupling: In this method, a functional group, which is attached to the surface, covalently interacts with a functional group from the target organic molecules (Fig. 5.3). The covalent bond can be formed either in a single step or multistep processes. Many methods exist for covalently attaching one functional group to another.

Cross-links are generally covalent bonds that link one organic molecule to another (Fig. 5.4). Cross-links are formed by chemical reactions that are initiated, when a form of energy is applied (such as irradiation, heat, and pressure).

The most important features of covalent bonds are that they are strong and selective. Their interactions are relatively fast and very well controlled. However, they are generally complex and might require extra processing steps, which can be costly and time consuming. Covalent bonds are strong but change the molecules, which may degrade the material biocompatibility. Cross-linking has similar advantages and disadvantages to covalent bonding. However, it provides more flexibility in establishing an interaction.

5.2.2 Self-Assembly

Self-assembly is a process in which chemical components such as atoms, molecules, and other building blocks self-assemble onto selected areas of functional systems. Self-assembly is normally self-driven by the energies of such systems [2]. In self-assembly processes the initial system is generally made of disordered components, which eventually form more organized patterns via non-covalent interactions. In nature, many biological systems use this process to form bio-structures that are generally rather more complex than their base units. By replicating the strategies that occur in nature, we can use self-assembly to modify the surfaces and sensitive layers of organic sensors and create novel chemical systems such as supramolecules.

Although the concepts of self-assembly are applicable to any material, currently the most promising avenues for self-assembly are those concerned with organic compounds. Such self-assembled materials are used in controlling the growth of organic materials for applications such as making sensitive and selective layers, developing superlattices of organic compounds with accurate thicknesses, and creating other components of the sensors such as insulating coatings. The self-assembly role in forming complex systems such as proteins (which will be explained later in this chapter) is of utmost importance. It takes place via *intermolecular* or *interamolecular* processes, such as folding, that allow the molecules adopt intelligent arrangements, which are capable of conducting desired functions, which are different from those of their subunits.

Self-assembled monolayers (*SAMs*) encompass a popular category of self-assembly systems that are used for forming functional or insulating layers. They are surfaces that consist of a single (mono) layer of molecules. SAMs are becoming increasingly accepted in different technologies, especially in sensors and accurate micro-/nano-device fabrication. This is due to the fact that they allow the film thickness and the composition to be precisely controlled at the scales as small as 0.1 nm. Many SAMs consist of *amphiphilic molecules*, which are molecules that contain both hydrophilic and hydrophobic groups. Due to the hydrophobic–hydrophilic nature of amphiphilic molecules, one side can show affinity to the surface and the other side to the component of the environment. As a result, they can arrange themselves in high orders onto many different surfaces. In addition to sensing, amphiphilic molecules are also used in a large number of applications such as in paint dispersions, cosmetic ingredients, detergents, and soaps.

Very often, in sensors amphiphilic molecules are employed to form surfaces suitable for molecular immobilization on transducers. In this case, one side of the molecules sticks to the sensors' surface and the other interacts with the biomolecules. Hence, they can potentially form a strong link between the target biomolecules and sensor's surface.

For establishing SAMs, a substrate is generally immersed into a dilute solution of the self-assembly molecules and a monolayer film gradually forms (Fig. 5.5). The monolayer formation can take just a few minutes to several hours.

In sensing applications, the effect of SAMs can be used directly. On the surface of acoustic wave transducers, SAMs formation can cause stress, which can affect

Fig. 5.5 Schematic depicting
the formation of SAMs
and the stages in the process:
(**a**) placing the solution
of SAMs' subunits on a
substrate, (**b**) adsorption,
and (**c**) organization

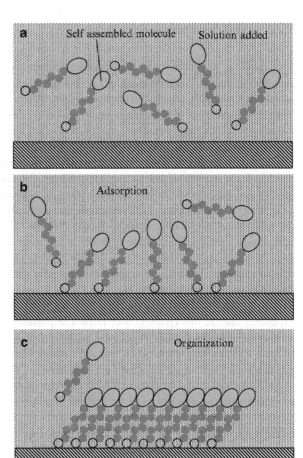

the phase velocity and amplitude of the propagating wave. SAMs can also be
fluorescently tagged and used in optical sensing. In conductometric sensors, they
can change the conductivity of the active surface, and their surface refraction
properties can be implemented in surface plasmon sensing. Additionally, they are
capable of altering the properties of double layers in electrochemical sensors.

By mixing self-assemble molecules with different end groups in the initial
preparation solution; we can produce mixed SAMs, which give interesting sensing
properties to the sensitive layers: one end group sensitive to one target analyte and
the other end group to a totally different target analyte. This produces sensitive
layers that can be used for detecting different targets. In addition, SAMs with no
active end groups or different chain lengths can be used as spacers (Fig. 5.6). These
spacers are sometimes necessary when we intend to sense large dimension
molecules to assure that no empty sites is left on the surface in order to avoid
nonspecific binding; hence nonspecific response by a sensor.

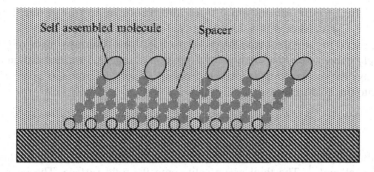

Fig 5.6 SAMs with a mixture of two different self-assembled molecules: one to produce the sensitive site and one as the spacer to cover the empty spaces

5.3 Surface Modification for Biosensing

There are many materials that a biosensor's surface can be made from. These materials can be inorganic such as gold and SiO_2 or organic such as polymers. The functionality of the surface will eventually depends on these coating materials. Some of the most common coatings, which are widely employed for forming the surface of biosensors, will be presented in this section.

5.3.1 Gold and Other Metallic Surfaces

Gold has a great compatibility with microfabrication industry standards. The deposition of gold thin films and utilization of gold surfaces are mature processes, which are widely implemented in the electronic industries. Gold is a noble metal that is commonly used in a wide range of applications in sensors. It does not oxidize easily so it can be reliably employed for making long lasting electrodes and also forming sensitive surfaces that do not change their properties in aqueous solutions. It can be deposited on the active area of a sensor by methods such as evaporation, sputtering, chemical processes, and electrodeposition. Gold itself does not make strong bonds to surfaces such as glass and silicon so to produce a strong gold adhesion to these surfaces, chromium or titanium intermediate layers are needed. Freshly deposited gold is quite hydrophilic. However, it rapidly adsorbs organic molecules and consequently becomes hydrophobic.

Gold is a suitable substrate for forming SAMs. The phenomenon of self-assembly on gold was first introduced by Nuzzo and Allara in early 1980s [3]. A common example is the formation of *thiol SAMs* on gold. A thiol is a compound that has a functional group composed of sulfur and hydrogen atoms ($-SH$). Alkane-thiols are among the most used thiols in sensing applications. An alkane-thiol

($HS-C_nH_{2n+1}$) is an alkane (which is a hydrocarbon that has the general formula of $-C_nH_{2n+2}$) together with a thiol functional group.

In general to form thiol type SAMs on the transducer's surface, made of gold, it is simply immersed in the solution for a duration of time and washed subsequently to remove the loose and unbonded components. Thiols de-protonate upon adsorption onto the gold surface as:

$$R - SH + Au \rightarrow R - S - Au + e^- + H^+. \tag{5.1}$$

Sulfur has a particular affinity to gold with a binding energy of approximately 20–35 kcal mol^{-1}. The thiol group splits onto the gold surface. The interaction creates a strong thiolate bond with the surface. Thiols assemble in the form of dense monolayers with a 2D order. With time, the SAMs undergo maturation and reorganization, which produces a more perfect and well-orientated layer. The alkane tails stand extended from the surface at an angle of ~30° normal to the substrate, and molecules are packed due to van der Waals forces (as depicted in Fig. 5.5c). Thiols ($R-SH$), sulfides ($R-S-R$), and disulfides ($R-S-S-R$) can all self-assemble onto gold.

Other metals such as silver (Ag), platinum (Pt), and copper (Cu) can also be used in the formation of SAMs. However, gold is the most popular choice, as Ag and Cu oxidize rapidly and Pt is not the most common material in microfabrication. In addition, Pt is a much more expensive material than gold and does not show particularly interesting Plasmon responses. Because of widespread interest in SAM technology, many different types of thiols with different chain length and end group functionalities are commercially available.

5.3.2 Silicon, Silicon Dioxide, and Metal Oxides Surfaces

Silicon (Si), *silicon dioxide (silica or SiO_2)*, and many other metal oxides establish the main backbone of the semiconductor and sensor industries. They are well studied and investigated materials, their properties are known, and there are standard tools for the fabrication of devices of various configurations based on them. Almost 100 %, the electronic industry has been built on silicon. It is a semiconductor with an intrinsic bandgap of 1.1 eV and electron mobility in the range of 1,000–1,500 cm^2 Vs^{-1} suitable for making electronic components such as transistors, solar cells, temperature sensors, mechanical (cantilever type) sensors, Hall effect sensors, photodiodes, and phototransistors. Silicon is also widely used in making electrochemical sensors such as ISFETS (Chap. 3). Silicon is not transparent to visible light due to its low bandgap; however, it is transparent in extended regions of the infrared spectrum.

Fig. 5.7 Oxygen-bridged Si terminated with hydroxyl groups on the surface of SiO$_2$ or Si

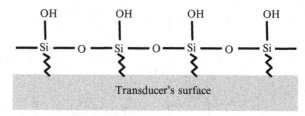

The insolating material of choice in semiconductor industry is SiO$_2$ (although high dielectric materials such as *hafnium oxide* (*HfO$_2$*) are rapidly taking its place). Depending on the synthesis process, SiO$_2$ can be amorphous such as glass or highly crystalline such as quartz. Different forms of silicon dioxide (such as silica, quartz, and glass) are transparent for visible light (glass) or both visible and UV light (quartz), and are both widely used in optical measurements. Many sensors rely on light absorbance, used within visual microscopy tools, and also for fluorescence measurements. In these cases, it is important that the substrate is transparent so that it does not interfere with the measurements. Piezoelectric properties of quartz is making it very popular for making acoustic wave based sensors. In addition, SiO$_2$, mixed with or doped by other metals and metal oxides, is used as a popular material in ion-selective membranes (Chap. 4).

It is possible to generate oxygen-bridged metal atoms and hydroxyl ($-$OH) groups on the surface of SiO$_2$ (Fig. 5.7). This can be easily done by exposing the surface to a mild acid. The *hydroxyl groups* make the surface quite hydrophilic. This process is also valid for the surface of silicon, as it very quickly forms a thin layer of oxide when exposed to mild acids or when even oxygen in air.

The formation of hydroxyl group also provides the opportunity to modify the surface with compounds such as *silane coupling agents* (Si-hydrocarbyl derivatives) . In the presence of water, highly reactive *silanols* that begin to condense, forming oligomeric structures, are produced (Fig. 5.8). Further condensation and dehydration between the coupling agent and the surface result in multiple strong, stable, covalent bonds on the surface. Silane coupling agents are commercially available with many different terminal function groups such as amine ($-$NH$_2$), thiol ($-$SH), and chloride ($-$Cl). These terminal groups can be used for coupling to organic molecules such as proteins and DNA. These attributes make them very attractive for the modification of Si- or SiO$_2$-based sensor surfaces compatible with silicon industry.

There are also many metal oxides, which are commonly used in sensing industries that can offer a variety of capabilities. Metal oxides such as *ZnO*, *TiO$_2$*, *SnO$_2$*, *WO$_3$* and *MoO$_3$* have large bandgaps (>2.5 eV), so they are mainly transparent in the visible and UV light ranges. Some of these metal oxides, such as WO$_3$ and MoO$_3$, can change color upon ionic intercalation. For instance, after exposure to H$^+$, WO$_3$ forms HWO$_3$ with a reduced bandgap material that is not transparent anymore. Such a process is the base for many *electrochromic* or *gasochromic* sensors. Additionally, smart windows also use such structures.

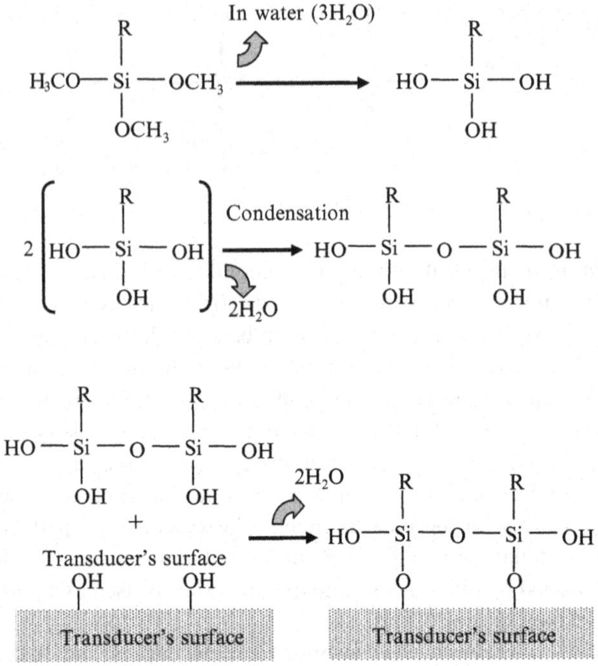

Fig. 5.8 Application of a silane coupling agent to produce a modified surface on SiO_2

Many metal oxides change their conductivity upon exposure to oxidizing or reducing gases or vapors. SnO_2 is commonly used as a *gas selective material* for gas species such as H_2, O_3, and also different *volatile organic compounds* (VOCs). Such metal oxides generally interact with selected gases at elevated temperatures. As a result, the metal oxide based gas sensors are generally accompanied by micro-heaters in their structures.

5.3.3   Carbon Surfaces

Carbon is an interesting material. Carbon atoms are capable of forming complicated networks, which are the basis of organic chemistry and biomaterials. Elemental carbon can form a number of very different structures. This includes well-known structures ranging from diamond and graphite, which have been known since ancient times, to fullerenes and nanotubes and more recently graphene (Fig. 5.9). The physical properties of carbon vary depending on its allotropic form: diamond is a very hard material with large bandgap (5.5 eV) that makes a great insulator. Refractive index of diamond is also comparatively large (approximately $n = 2.42$). Conversely, graphite is opaque and black. Graphite is a layered material made of

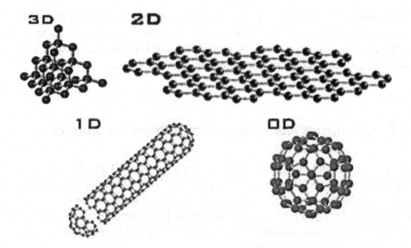

Fig. 5.9 Crystal structures of the different allotropes of carbon. Three-dimensional diamond and graphite (3D); two-dimensional graphene (2D); one-dimensional nanotubes (1D); and zero-dimensional buckyballs (0D) (adapted and reprinted with permission from [4])

planes that are loosely bond together by van der Waals forces and can be easily separated from one another. Graphite is also highly electrically conductive along the direction of the planes.

Carbon base materials have been increasingly favorable for incorporation in sensors after the emergence of carbon nanotubes and graphene. Both graphene and carbon nanotubes are highly electrically and thermally conductive along the direction of the planes and tubes, respectively, allowing electric charges and heat to move along them with minimal resistance. These properties are extremely beneficial for developing highly sensitive sensors in which charges or heat should be conducted freely, within the body of the sensor, to minimize the reading losses. Graphene and carbon nanotubes are also mechanically hard along these tubes and planes, making them amongst the strongest materials for the development of hardly breakable and, at the same time, light structures. This property is of utmost importance for the development of cantilever type sensors and the type of tips that are used in atomic force microscopes that require small tips for better imaging resolutions.

Both graphene and carbon nanotubes can provide large surface areas which are essential for the increasing surface to volume ratio of a sensor's sensitive layer that allows larger responses. They are also capable of functioning as point sources that increase the local electric field (which interact with double layers where the ion targets are located—as described in Chap. 4) dramatically, hence increasing the sensitivity of such devices.

Carbon is widely used as electrodes for electrochemical sensors. Fresh carbon surfaces are very hydrophobic, which can adsorb many organic species. This increases the nonspecific binding, which reduces the selectivity and sensitivity of the sensing system but at the same time provides great opportunities in functionalizing its surface via different methods. Due to great interests in carbon-based materials, a large number of procedures, for functionalizing such surfaces, have been devised and are now available in public domain.

5.3.4   Conductive and Nonconductive Polymeric Surfaces

Polymers are large organic molecules that consist of repeating units, called monomers, which are covalently connected. They are attractive materials for the development of sensors as their chemical and physical properties can be tailored over a wide range. Some of their advantages include: low fabrication costs, ability to take different forms, possible biocompatibility, and their ability for chemical and biosensing at different temperatures.

Both *intrinsically conducting* and *nonconducting polymers* (*ICPs* and *NCPs*) can be used for developing sensors. They can be used in sensor structure, directly participate in sensing a sensitive layer, and can also be used as the media for immobilizing biomaterials.

5.3.4.1   Nonconducting Polymers

NCPs are becoming increasingly attractive in the fabrication of transducers and sensors. They are inexpensive, simple to prepare, and it is relatively easy to manipulate their shape. Screen printing, molding, and stamping are generally employed to carry out such tasks. With a plethora of photo and thermally curable NCPs available, there are many possibilities that they can be incorporated into sensors' fabrication and design. NCPs can also be employed as selective layers in bio- and ion-selective sensing, with sensitivities that can be manipulated using different fillers and surface chemistries.

NCPs have many applications in biosensing. They can be used for the entrapment of biomolecules and as membranes for the formation of bio-selective layers for functionalizing surfaces for the immobilization of biomolecules such as proteins and DNA.

Surface of polymers can be extremely diverse, ranging from extremely hydrophobic to extremely hydrophilic. For instance, *polysaccharides* such as *cellulose* and *agarose* are very hydrophilic due to the dominance of hydroxyl ($-OH$) groups. These are well suited for protein adsorption as the proteins do not denaturing on these surfaces (denaturing process will be explained in later sections).

As another example, polymers, such as *carboxymethyl dextran* (cosmetic ingredient for hair care and skin care products), can efficiently increase the concentration of ion exchange groups near to the surface. Such surfaces are now widely used in electrochemical and surface Plasmon type surfaces.

a   **b**

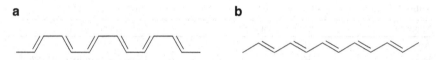

Fig. 5.10 Conjugated structure of polyacetylene: (**a**) *cis* and (**b**) *trans*

In many cases, biomaterials do not directly adsorb onto different types of polymeric surfaces. As a result, biomaterials such as proteins need to be immobilized covalently. However, hydroxyl groups are difficult to be employed for direct covalent bonding onto many types of polymers. Therefore, we need to make surfaces more reactive using activating reagents, which make them ready for direct coupling. Many methods can be found in literature for the manipulation of polymeric surfaces [5].

5.3.4.2   Intrinsically Conducting Polymers

ICPs (or *synthetic metals*) are polymers that can show electrical, magnetic, and optical properties similar to those of metals and semiconductors. In 2000 Alan J. Heegar (USA), Alan McDiarmids (New Zealand), and Hideki Shirakawa (Japan) received the Noble prize in Chemistry for their contribution in the discovery of ICPs in the mid-1970s.

These days, ICPs such as *polyaniline*, *polythiophene*, *polypyrrole*, and *polyacetylene* are increasingly play more significant roles in the development of sensors. They are attractive materials for sensor development as their electrical, mechanical, and optical properties change, when exposed to different target analytes as well as physical stimuli.

For a polymer to be conducting it has to alternate single and double bonds along the backbone of the polymer, which is called *conjugation*, and the resultant polymer is described as *conjugated* (Fig. 5.10).

Table 5.1 shows the degree of conductivity of different forms of polyaniline in comparison with conventional conductive and nonconductive materials.

When dealing with ICPs for sensing applications, doping is an important process as the users can tailor the conductivity of polymer to suit their needs. The conductivity of ICPs can be manipulated by doping them with certain atoms or molecules.

5.4   Proteins Incorporated in Sensing Platforms

Proteins are biochemical compounds that carry out biological functions. They perform many tasks within live tissues, and hence such functionalities can be borrowed and incorporated into biosensors. Their use as sensitive materials, pores

Table 5.1 Conductivity of polyaniline in comparison to some common materials

Material	Conductivity (S/m)	Conductive polymer
Ag	10^6	
In	10^3	Doped-polyaniline 10^5 (S/m)
Ge	1	
Si	10^{-6}	
Glass	10^{-9}	Intrinsic polyaniline 10^{-10} (S/m)
Diamond	10^{-12}	
Quartz	10^{-15}	

with triggers, switches, self-assembling arrays, and motors are just a few niche applications, for which proteins can be employed.

Proteins can perform a large number of functions, which are essential for living creatures. For instance, the ion channels in cell membranes, which are made of proteins, can keep the ion balance across our cells by letting single ions passage. Proteins, which operate as nanosize motors, can use energy from small molecules to throw themselves forward in order to walk on a surface within a cell. Enzymes can catalyze a large number of chemical interactions essential for maintaining operation of a living creature. There are many more.

In this section, structures of proteins will be briefly described and their most important sensing applications will be presented.

5.4.1 Structure of Proteins

Each living organism has a certain number of protein types. The total number is different in various organisms. For instance, yeast has approximately 6,000 proteins but the number approximately exceeds 32,000 in human beings [6]. *Amino acids* are the basic subunits of proteins. A simple amino acid, alanine, is shown Fig. 5.11. Alanine consists of a carbon, which is bound onto an amino group (–NH$_2$), a carboxyl group (–COOH), and a methyl side chain (–CH$_3$).

All amino acids possess a carboxylic acid group and an amino group, both linked to the same carbon, which is called the α-carbon. The chemical variety in amino acids comes from the side chains (−R), which is also attached to the α-carbon. In alanine the side chain is –CH$_3$.

The covalent bond between two adjacent amino acids is called a *peptide bond* (Fig. 5.12). The chain of amino acid is called a *polypeptide*. The combination of amine in a carboxyl group is given the polypeptide directionality.

There are 20 different types of amino acids that are commonly found in proteins, each with different side chain (−R in Fig. 5.12) attached to the α-carbon. One of the mysteries of evolution is that only 20 different amino acids appear over and over again in all proteins, whether they are from humans or from bacteria. These

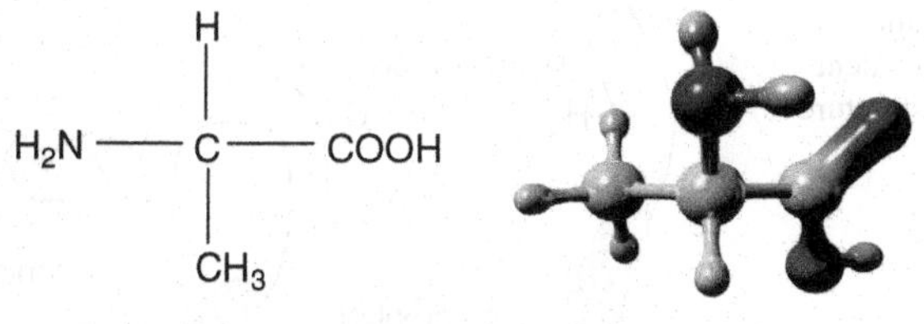

Fig. 5.11   The structure of an alanine molecule

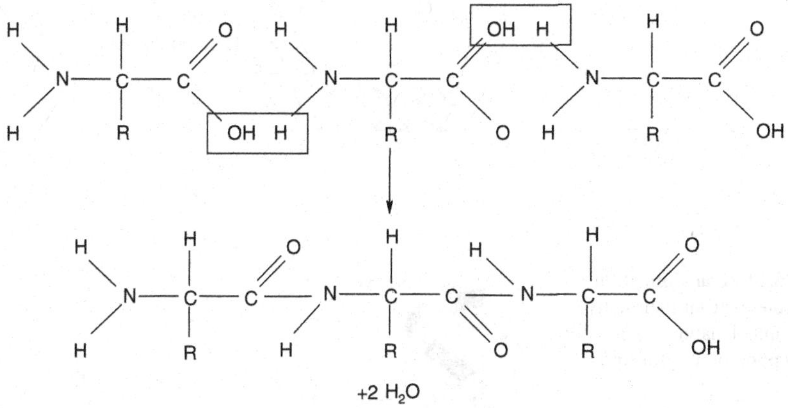

Fig. 5.12   The joining of three amino acids through peptide bonds, which forms a polypeptide chain

20 standard amino acids provide a remarkable chemical versatility to proteins. Five of these amino acids can ionize in aqueous solutions, the other remain unchanged. Some amino acids are polar and hydrophilic and some are nonpolar and hydrophobic.

A protein is made from long chains of amino acids. As such, proteins are polypeptides. Each type of protein has a unique sequence of amino acids. Long polypeptides are very flexible. Many of the covalent bonds that link the atoms in an extended chain of amino acids allow the free rotation of the atoms they join. As a result, proteins can fold in an enormous number of ways. Once folded, each chain is restrained by non-covalent bonds. These non-covalent bonds, which maintain the protein shape, include hydrogen bonds, ionic bonds, and van der Waals forces (Fig. 5.13).

Proteins fold into a conformation of lowest energy in order to minimize their free energy. These conformations are the most stable states in which the proteins exist. Nevertheless, this stable condition can change when a protein interacts with the other molecules in a cell. When a protein folds improperly, it can form aggregates that can damage cells and even whole tissues. Alzheimer's and Mad Cow diseases are caused by the aggregation of proteins. In living organisms, generally the protein folding process is assisted by special proteins called *molecular chaperons*. Proteins

Fig. 5.13 Schematic
example of non-covalent
bonds within the structure
of a protein

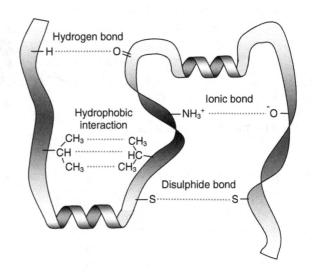

Fig. 5.14 Transformation
of proteins from their native
state into denatured state after
adsorption onto a substrate

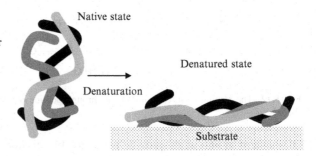

can also be unfolded or denatured by certain solvents that can disrupt their non-covalent interactions (Fig. 5.14).

Proteins can be directly adsorbed onto the surface of a sensor; however, in the process they may become *denatured* (lose their initial structure—Fig. 5.14). After denaturing, a protein may lose its functionality, which can be a major problem in sensing applications. For example, an antibody (which will be explained later) must retain its shape in order to retain its sensitive properties.

Proteins are the most diverse macromolecules in a cell. Their size can range from approximately 30 to 10,000 amino acids, they can be globular or fibrous, and they can form filaments, sheets, rings, or spheres.

Revealing the *amino acid sequence* is an important part of analyzing a protein and its functionalities. The first protein sequenced was beef insulin by Fred Sanger, the 1958 Nobel Prize winner from the University of Cambridge. Generally, the analysis of a protein begins by determining its amino acid sequence. The cells are broken, open, and the proteins are separated and purified so as to conduct a direct analysis on the chemical components. Alternatively, the genes that encode proteins can be sequenced instead. Once the order of the nucleotides in the DNA, which

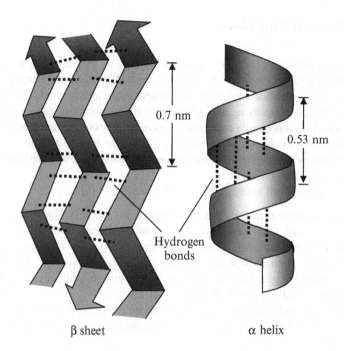

Fig. 5.15 Representations of α-helix and the β-sheet folding patterns

encodes proteins, is known, the information can be converted into the corresponding amino acid sequence. Generally a combination of these direct and indirect methods is used for a comprehensive analysis of a protein.

The α-*helix* and the β-*sheet* are common folding patterns for proteins (Fig. 5.15) and they are found in the structure of most proteins. The α-helix was first found in a protein called α-*keratin*, which is abundant in skin. α-helix resembles a spiral staircase as within its structure a single polypeptide chain turns around itself to form a structurally rigid cylinder.

The β-sheet is the second folded structure, which was first found in protein *fibroin* that is major constituent of silk. β-sheets form rigid structures at the core of many proteins and perform functions such as giving silk fibers their tensile strength and keeping insects from freezing in the cold.

There are many higher levels of organizations for proteins rather than just α-helix and β-sheets. The order in which amino acids are linked gives the *primary structure* of the protein. The next level up is the *secondary*, which is concerned with the shape of the protein in localized areas and takes into account the α-helix and β-sheets that form within certain segments of the polypeptide chain. The full three-dimensional protein structure is the *tertiary structure* and it consists of all α-helix and β-sheets, coils, loops, and folds in the protein. The *quaternary structure* of a protein is the overall structure in the case that the protein comprises more polypeptide chain. An example of this is *Hemoglobin*, which has four polypeptide chains in its quaternary structure.

5.4.2 Proteins Functions

Proteins represent an exciting area in biotechnology as they have ideal properties for engineering tasks such as processing sophisticated architectures at very low (nanoscale) dimensions, have rich chemistry, and provide us versatile functionalities. By employing the knowledge of biochemical engineering, it is possible to harness a protein's power so as to create new components for intelligent systems and devices. It is also possible to synthesize basic artificial proteins of desired sequences, which can function as sensing elements, drugs, and small smart complexes such as *nanorobots*.

Proteins can perform many different tasks as follows [6]:

- Generate movements in cells and tissues via motor proteins: *Myosin* and *Kinesin* (Fig. 5.16) are such proteins. For instance, Kinesin can attach itself to microtubules and surfaces and moves along them in order to transport cellular cargo holders, such as vesicles. It uses *adenosine triphosphate (ATP)* as the energy source to generate the force that it needs to walk along. *Adenosine diphosphate (ADP)* is the by-product.
- Transport materials, such as small molecules and ions: for example, in the blood stream *Serum Albumin* carries lipids, *Hemoglobin* carries oxygen, and *Trasferrin* carries iron.
- Receiver of signals: some proteins can detect signals and send them to the cell's response machinery. *Rhodopsin* in retina, for example, detects light.
- Storage of small molecules or ions: for example, iron is stored in the liver within the protein *Ferritin*.
- Promote intermolecular chemical interactions: *enzymes* generally catalyze the breakage and formation of covalent bonds, as in *DNA polymerase*, which is used in copying DNA.
- Function as selective valves: for instance, proteins embedded in the plasma membranes form channels and pumps that control the passage of nutrients and other small molecules into and out of cell.
- Carry messages from one cell to another: for example, *Insulin* is a small protein that controls the glucose levels in blood.
- Serve as nanomolecular mechanical machines: for example, *Topoisomerase* can be used for untangling DNA molecules.
- Act as antibodies.
- Perform as toxins.
- Act as antifreeze molecules: for instance, some proteins in arctic fish can help them to avoid freezing in colder waters.
- Form elastic fibers.
- Generate light through luminescent reactions.
- Form glues: The glue that forms in mussels is a protein, which is produced allowing a mussel to attach to other objects and many more.

Fig. 5.16 How Kinesin
functions as a motor protein

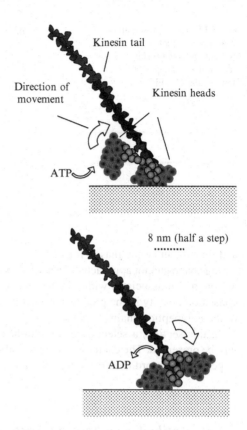

5.4.3   Proteins in Sensing Applications

Generally, the conformation of proteins gives them their unique functionalities and how they interact with chemical and biochemical components in the environment. The binding of proteins to other chemicals is not always strong; in many cases it is actually weak. However, the binding always shows specificity, which means that each protein can only bind to one, or at most, only a few selected chemical components that it encounters. Such a nature can be efficiently used for making sensitive surfaces.

The substance, which is bound by a protein, is called a *ligand* for that protein. This ligand can be an ion, a small molecule, or a macromolecule [6]. *Ligand–protein binding* is of great importance in sensor technology. Non-covalent bonds such as hydrogen bonds, ionic bonds, and the Van der Waals forces as well as hydrophobic interactions are responsible for the ability of selective binding of a protein to a ligand. The effect of each bond can be weak but the simultaneous formation of many weak bonds between the protein and the ligand form a rather strong selective binding. Even the matching of the surface contour of the protein

Fig. 5.17 The representation
of a protein–ligand
interaction as the result
of a number of separate
non-covalent bonds

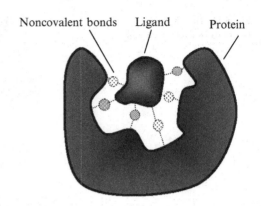

Noncovalent bonds   Ligand   Protein

and ligand can be the reason for selective binding (Fig. 5.17), as molecules of the wrong shape cannot approach the active sites closely enough to bind. The region of a protein that associates with a ligand is called a *binding site*. These sites usually consist of a cavity in the protein surface that is formed by the folding arrangement of the polypeptide chains.

In this section, a selection of protein-based components in sensor technology will be presented. The chemical and physical properties of a protein give the ability to perform extraordinary dynamic processes.

5.4.4 Antibodies in Sensing Applications

Antibodies (belong to *immunoglobulins* family) are proteins which are produced by the immune system in response to foreign materials [6]. The most common of these foreign molecules are invading *microorganisms* such as *viruses* and *bacteria*. Antibodies either render them inactive or prepare them for destruction. The proteins of the antibody family are highly developed with the capacity to bind to particular ligands. The antibody family consists of five subdivisions: *IgG, IgM, IgA, IgE*, and *IgD*. An antibody recognizes the target, which is generally called an *antigen*, with a high specificity. Antibodies are Y-shaped molecules (as seen Fig. 5.18). It has been identified that antibodies have two identical binding sites. These binding sites can conform to a small portion of the antigen. The antibody binding sites are formed from several loops of polypeptide chains, which are connected to the end of the *protein domains*. These protein domains are made of four polypeptide chains, two of which are identical heavy chains and the other two are identical light chains. These chains are all held together by disulfide bounds [6].

The amino acids in the binding sites can be changed by *mutation* without altering the domain structure of the antibody. A large number of antibody binding sites can be formed by changing the length and the amino acid sequence of the loops.

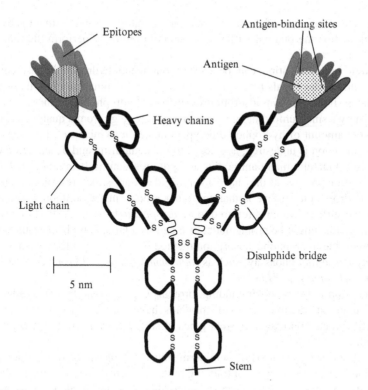

Fig. 5.18 The general structure of an antibody

The reactivity of an antigen is restricted to specific parts of the molecules which are called *epitopes*. These epitopes specifically bind to the antibody binding sites, which are also known as *paratopes*. The area of the antigen epitopes surface, comprising between 15 and 22 amino acids, interacts with the same number of amino acids on the antibody paratopes. Many intermolecular hydrogen bonds, ionic bonds, and salt bridges participate in these interactions and apparently water molecules also participate in these interactions.

When suspended in water the antibody and antigen protein molecules initially repel each other due to their hydrophilic nature. This force must be overcome before the epitopes and paratopes are brought together. The major reasons for the attractions are probably electrostatic interactions and still several available hydrophobic sites on antibodies and antigens.

For developing biosensors, which are sensitive to specific antibodies or antigens, their corresponding antibodies or antigens should be either immobilized (anchored to the surface) on the active area of affinity-type sensors or embedded within the bulk of bulk-type sensors.

For successful antibody or antigen immobilizations, the functional sites of the proteins should be recognized. Consequently, a procedure should be developed to realize the binding of the proteins on the surface, while keeping them active. In the immobilization of antibodies the epitopes and paratopes should remain active and available, while the other site (the *stem*) should preferably perform the immobilization task.

Protein immobilization can be carried out using hydrogen bonds, covalent bonds, via van der Waals forces, ionic bonds, or a combination of all these bonds and forces. If the protein adsorption occur on a hydrophobic surface, it is often followed by a small unfolding of a protein structure (e.g., from a quaternary, which causes the amount of hydrophobic polypeptide chain to increase). In many cases, this phenomenon is undesired, as it leads to the denaturing and deactivation of the protein. Fortunately, many antibodies are quite resistant to being unfolded during the adsorption process, as they have a very rigid tertiary structure. Another problem with a hydrophobic surface is that any other protein in the solution also tends to bind to the surface, which is referred to as *nonspecific binding* (*NSB*). For developing sensors this might be problematic, as it causes a large extra background signal. Altogether, the hydrophobic adsorption method is very popular for well-developed antibody-based immunoassays such as *radioimmunoassay* (*RIA*) and *enzyme-linked immunosorbent assay* (*ELISA*).

In addition to hydrophobic binding, proteins can also bind to a charged surface due to ionic interactions, as some proteins have ionized surface groups. This interaction occurs on many different types of surfaces, even those that are weakly charged.

As described previously, semipermeable polymeric and non-polymeric membranes can be employed for trapping proteins, such as enzymes and antibodies. Generally the proteins which are captured this way are small proteins, having molecular weights less than 10 kDa (kilo Dalton—Dalton is an atomic mass unit and has a value of 1.66×10^{-27} kg). Several well-studied membranes include polycarbonate and nylon. Proteins may also become physically entrapped within the volume of a hydrogel. A hydrogel can be made of a polymer, which is dissolvable in warm water and gels when cooled. Gelling is due to hydrogen bonding. The most widely used hydrogel polymer is *agarose*, which gels at approximately 40 °C. It is highly porous, which makes it more suited to microorganisms and organelles as opposed to single proteins.

There are many biological assays involving antibody–antigen interactions which are used in sensing procedures. The most common of these assays are *direct assay* and *competitive assays*. In direct assays, the antibodies or other receptor molecules are immobilized onto a sensor's surface. The target molecules can then selectively bind to these molecules (Fig. 5.19). Direct assays find applications in affinity sensors such as piezoelectric, optical waveguide-based and surface plasmon sensors, where the evanescent waves penetrate into the added layer, producing a response. No optical tagging is needed in these sensors as they are able to sense the mass (in fact, the perturbation caused by the thickness) of the added layer.

In a competitive assay, firstly the antibody is immobilized on the substrate's surface. Then a known quantity of *labeled analytes*, which interact with the target molecules, is added and they compete with the unlabelled molecules for the available antibody bonding sites on the substrate (Fig. 5.20). The result can be

Fig. 5.19 A simplified
depiction of a direct assay

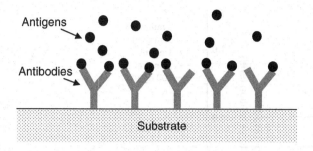

Fig. 5.20 A simplified
schematic of a competitive
assay

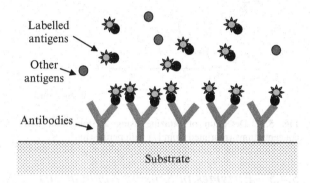

optically investigated and interpreted using the relationship that the higher the
optical beam (e.g., from a laser source) interact on the surface the more of target
analytes are detected.

In using antibodies or antigens as selective layers, the cross-sensitivity for with
other components in the environment is an important issue. In practical sensing
applications, the *specificity* is very important as it rates the sensor's level of
discrimination for recognizing a particular analyte amongst other species.
Antibodies with greater specificity also have a higher affinity to their complementary antigens.

In addition to piezoelectric and optical devices, antibody-based immunosensors
can also be based on electrochemical and amperometric transducers. In this case the
antibody–antigen reaction alters the surface charge or conductivity of the layer.
Surfaces, which are modified using antibodies and antigens, are also used for
pesticides, bacteria and virus sensing, as well as medical sensing applications
such as in drug recognition.

As an example, the output response of a typical antibody–antigen sensor is
presented in Fig. 5.21 [7]. In this example, a SPR biosensor is used for sensing a
pesticide (atrazine in this case). A sample that contains a mixture of atrazine
antibody is exposed to an atrazine derivative-coated SPR biosensor. As can be
seen, pesticide concentrations as low as 100 pg mL^{-1} of atrazine can be readily
sensed.

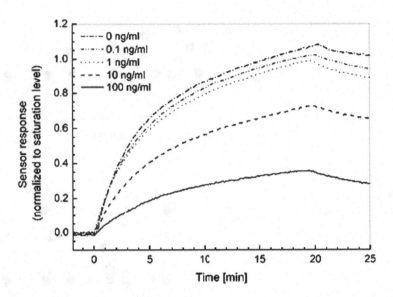

Fig. 5.21 Detection of atrazine using an inhibition assay at different atrazine antibody concentrations (reprinted from [7] with permission)

5.4.5 Enzymes in Sensing Applications

There are many other proteins that are different from antibodies, for which their attachment to a ligand is only the first step of their operation. This is the case for a large class of proteins, which are called *enzymes*. Enzymes bind to one or more ligands, which are called *substrates*, and convert them into chemically or physically modified products (Fig. 5.22).

Enzymes speed up chemical reactions without participating in them. As a result, they are categorized as *catalysts*, which are materials that accelerate chemical reactions without themselves undergoing a net change.

Each type of enzyme is highly specific, catalyzing only a single type of reaction. For instance, the blood clotting enzyme *Thrombin* cleaves a particular type of blood protein in a specific place and nowhere else. Enzymes often work in teams, where the product of one enzyme–substrate interaction can become the substrate of another enzyme.

There are many enzyme-based biological assays, which are used in sensing the presence of antigens. Assays such as *ELISA* are commonly used together with optical sensing systems. ELISA generally uses two antibodies: one antibody is specific to the antigen and the other antibody reacts with an antigen–antibody complex which is coupled to an enzyme. This second antibody, which is generally labeled with fluorescent or chromatic molecules, is used for producing a detectable signal. A *sandwich ELISA* procedure is shown in Fig. 5.23.

Fig. 5.22 The representation of an enzyme–substrate interaction

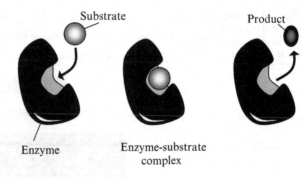

Substrate

Product

Enzyme

Enzyme-substrate complex

Fig. 5.23 A schematic of an ELISA. (1) The first antibody is immobilized on the substrate, (2) the target antigen interacts with the antibody, (3) the detecting antibody (second antibody) is added which binds to the other side of antigen. This secondary antibody is an enzyme-linked complex and (4) the substrate is added. It is the substrate–enzyme interaction, which produces a detectable optical signal

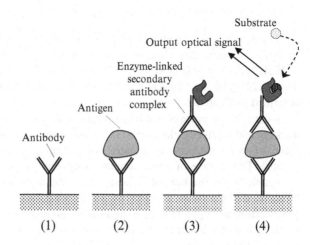

Substrate

Output optical signal

Enzyme-linked secondary antibody complex

Antigen

Antibody

(1) (2) (3) (4)

Example: *Blood Glucose Sensors*

One of the most important enzyme types in sensing applications are the *redox enzymes* groups. In redox reactions inside the body of a living organism a substrate becomes oxidized or reduced. This action uses energy and support life within the body of the living organism. Redox enzymes accelerate the redox reactions to render them biologically useful. Redox enzymes also control the highly reactive intermediates. As a result, in such enzymes generally only a single substrate is employed and a single specified product is produced.

Redox enzymes are used in many sensing applications such as the glucose sensors. However, these redox enzymes usually lack direct communications with electrodes. As a result, they have to be used along with materials, which mediate the diffusion of electrons into electrodes.

One of the most popular electrochemical sensors in the market is the blood glucose monitor. Patients with diabetes require regular monitoring of the glucose

Fig. 5.24 A schematic
of a working electrode
for the early type of glucose
electrochemical sensors

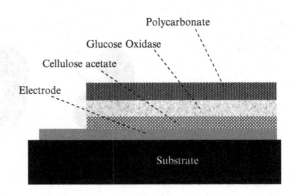

level in their blood. Sometimes, process has to be conducted several times a day to
allow them control their disease through insulin injections.

Early types of home glucose sensors consisted of a three-layer working electrode
(Fig. 5.24) and an Ag/AgCl reference electrode (reference electrodes were
described in Chap. 3).

In the sensing process, a drop of the patient's blood is placed on the surface of
working and reference electrodes. In the working electrode, the outer layer can be
made of *polycarbonate* (polymers containing carbonate $(-O-(C=O)-O-)$
groups). The surface of polycarbonate is hydrophilic and, more importantly, it is
permeable to glucose but not to most of the other constituents of blood. The middle
layer of the electrode consists of an enzyme, *glucose oxidase*, which forms
gluconolactone (a gluconic acid) and hydrogen peroxide (H_2O_2) after the reaction
with glucose. The underneath layer is made of cellulose acetate, which is permeable
to H_2O_2. The hydrogen peroxide is oxidized at the working electrode that is held at
a positive voltage (approximately +0.6 V) versus the Ag/AgCl reference electrode.
This produces a current according to the redox half-equation:

$$H_2O_2 \rightarrow O_2 + 2H^+ + 2e^-. \tag{5.2}$$

The resulting current is proportional to the glucose concentration of the sample,
and so the blood sugar levels can be monitored.

5.4.6 Transmembrane Sensors

The membrane of a cell contains a number of incorporated membrane proteins
(*transmembrane proteins*) . These proteins can be integrated within, placed on the
outside of, or throughout the membrane (Fig. 5.25).

A group of the *transmembrane proteins* consist of polypeptide chains, which
cross the *lipid bilayer* (the building blocks of the membrane structures) and usually

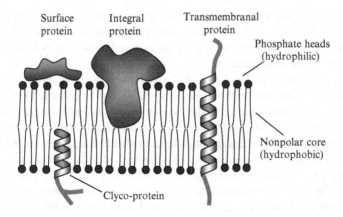

Fig. 5.25 A cross section of a membrane, showing the phospholipid bilayer and examples of some proteins associated with the membrane

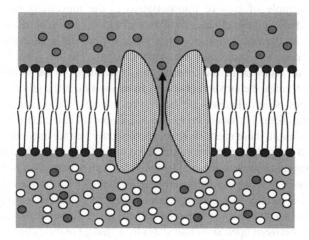

Fig. 5.26 A representation of a transmembrane transporter allowing the selective passage of particles

have an α-helix form (Fig. 5.26). In many transmembrane proteins the polypeptide chain crosses the membrane only once. Many of these types of proteins form receptors for the *extracellular signals*.

There is another type of proteins, which acts as *transmembrane transporters*. These proteins allow nutrients, metabolites, and ions to travel across the lipid bilayers (Fig. 5.26). These proteins are necessary as the lipid bilayers are highly impermeable to all ions and charged molecules as well as many nutrients and wastes such as sugars, amino acids, nucleotides, and many cell metabolites.

Each transport protein provides a private passageway across the membrane for a special class of molecules. These membrane proteins extend through the bilayer.

They have both hydrophobic and hydrophilic regions. The hydrophobic region is in the interior of the bilayer, which binds to the hydrophobic tails of the lipid bilayer. The hydrophilic regions of these proteins are exposed to the aqueous environment on either side of the membrane.

The transportation of molecules can be either passive or active. Protein is passive when it automatically, and without any control, allows the passage of molecules, generally when the concentration of the target molecule is higher on one side of the protein gate than the other side. However, if the transportation is against the concentration gradient then energy must be provided to the system to move these molecules to higher concentration site, which is termed as *active transport*.

One example of *passive transportation* is the transportation of glucose in the plasma membrane of mammalian liver cells. This transport protein can adopt two different conformations: in one conformation glucose binds to the exterior of the cell and in the other it is exposed to the interior of the cell. As glucose is an uncharged particle, so the mass concentration has to be measured and then signals must be sent to the valve proteins.

Another example of passive transportation is the movement of ions across membranes. Many cell membranes have an embedded voltage across them, which is referred to as the *membrane potential*. This difference in potential exerts an electrical force on any molecules that carry charges. The net force is called the *electrochemical gradient* of the membrane. Depending on the direction of this gradient, it may cause the movement of ions from one side to the other or vice versa.

Special types of ion channels are interesting examples of active transportation gates. These ion channels concern the exclusive transport of ions such as Na^+, K^+, Cl^-, and Ca^{2+}. As a result, these channels should be selective, letting only one type of ions to pass through and only in one direction. For example, for many of these pores, when ions pass through, a transient contact of ions with the wall of pores occurs. This contact allows the protein to recognize them and only let the right ions to pass. This contact also works as a counter that generates a passage rate count for a cell. These ion channels are not continuously open but instead are gated. A specific stimulus triggers them to switch between open and close states by a change in their conformation.

Membrane transporters are useful in the development of accurate and selective sensors. They are extremely selective and can respond to even a single molecule or ion. The ion channels described above are also called *nanopores*. Many examples of their applications as selective membranes can be found in the literature and there are many practical ways of using transporters. In the case of ion channels, it is possible to measure the current, which is generated by the passage of single ions through these channels.

One of the procedures for conducting such measurements is *patch-clamp recording*. Neher and Sakmann developed the patch clamp in the late 1970s and further perfected it in the early 1980s. They received the Nobel Prize in Physiology or Medicine in 1991 for this work. In a patch-clamp recording, a glass tube with an open tip, of a diameter of only a few micrometers is used (Fig. 5.27). The glass electrode is filled with an aqueous conductive solution. Then, this tip is pressed

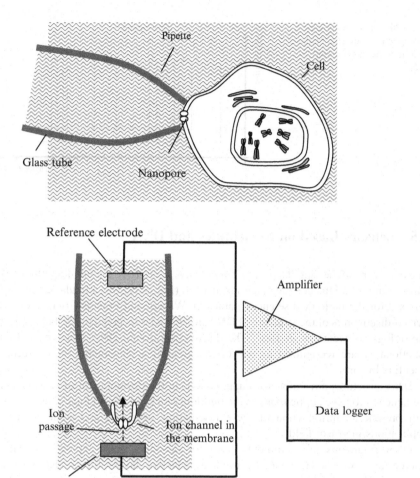

Fig. 5.27 The patch and clamp procedure (*above*) and the set up of a patch-clamp ion sensor (*below*)

against the wall of a cell membrane. With gentle suction a small part of this membrane can be removed. A metal wire is inserted at the other end of the tube. Finally, this *microtube electrode* can be used in a solution of ions along with other metal electrodes, together with accurate circuits, to measure the generated currents (Fig. 5.27).

The current measured can be in the range of pico-amps, which almost represents for the passage of single ions. Even if the conditions of measurements are kept constant, a random square wave signal is always recorded. This shows that the ion channels snap between their closed and opened states randomly (Fig. 5.28).

Fig. 5.28 The opening and closing of the pore change the current in the patch-clamp setup

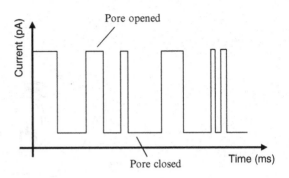

5.5   Sensors Based on Nucleotides and DNA

The discovery of the fact that *deoxyribonucleic acids* (*DNA*), the building blocks of chromosomes, are the genetic materials in a cell (Fig. 5.29) was a fundamental leap forward for the biological sciences. James D. Watson and Francis Crick were the two co-discoverers of the structure of DNA in 1953 who received the Nobel Prize in Physiology or Medicine in 1962. The discoveries around DNA structure and its functionality are recognized as one of the greatest achievements of our recent scientific history.

DNA structures have numerous applications in the development of sensors and are believed to have applications as the building blocks of DNA chips. This section will present an introduction of these fascinating structures and some of their applications in sensor field.

DNA fragments are commonly used for decoding gene expressions using microarray sensors. They can be used as selective layers for conductometric sensors. DNA fragments are readymade tools in biology for the construction of biomaterials. DNA has an excellent binary structure with embedded data for the selective detection of protein structures.

The ability of cells to *store*, *retrieve*, and *translate* the *genetic information* is what our lives depend upon. The combination of these activities maintains a *living organism* and differentiates it from other materials. At cell division, this hereditary information is passed on from a cell to its *daughter cell*. The information is stored in the *genes* within a cell (Fig. 5.29), which are information-containing elements for the synthesis of particular proteins. As such, they determine the characteristics of a species. The number of genes is approximately 30,000 for humans (equal to the number of proteins). The information in genes is copied and transmitted from cells to daughter cells again and again during the life of a multicellular organism, which is carried out with a supreme accuracy. This process keeps the genetic *code* fundamentally unchanged during the life of an organism.

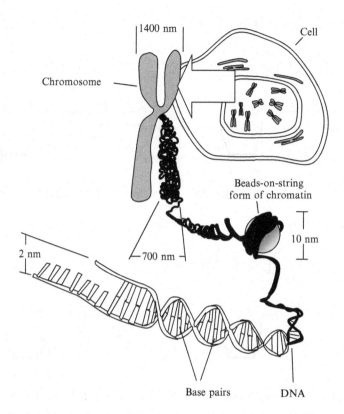

Fig. 5.29 A schematic showing the relationship between chromosome and DNA

5.5.1 The Structure of DNA

The main building blocks for DNA are *nucleotides* that are composed of: (1) a *phosphate*, (2) a *sugar*, and (3) a *base* [6]. The nucleotides are covalently linked together into a *polynucleotide chain*. The *sugar–phosphate* chain is the backbone from which the bases are extended. The four bases are *adenine* (*A*), *cytosine* (*C*), *guanine* (*G*), and *thymine* (*T*) (Fig. 5.30).

In early 1950, the examination of DNA using X-ray diffraction revealed that it is composed of a helix of two twisted strands (Fig. 5.31). The model proposed by Crick and Watson described the possibility of using such a helix for encoding and replication of proteins. A DNA molecule consists of two long chains known as *DNA strands*. These strands are composed of four types of *nucleotide subunits* and they are held together by *hydrogen bonds*.

Nucleotide subunits are linked together at a directional way (Fig. 5.31) and as a result, the DNA strands have polarity. At the ends of chains are the *3′ hydroxyl* and the *5′ phosphate* groups, which are referred as the *3′ end* and the *5′ end*.

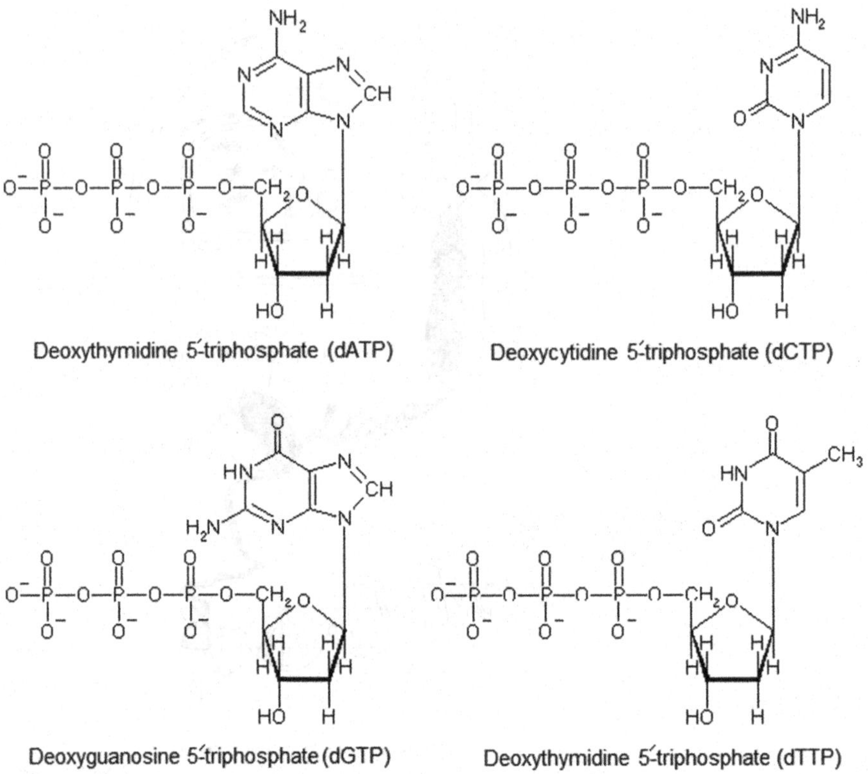

Fig. 5.30 The four DNA nucleotides

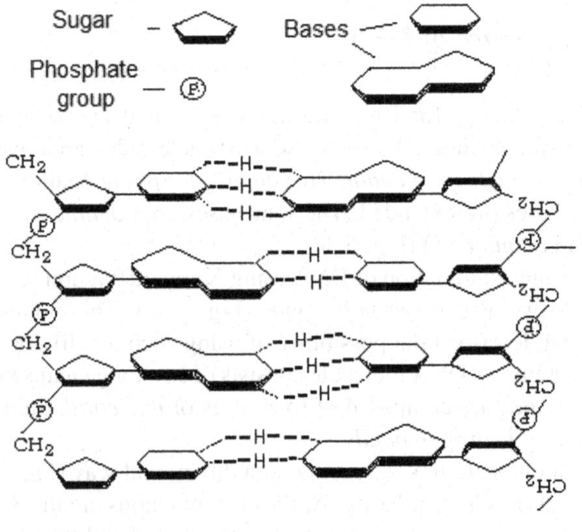

Fig. 5.31 The structure of DNA strands

The bases are paired by hydrogen bonds: "A" selectively pairs with "T" and "C" selectively pairs with "G." The two sugar–phosphate backbones twist around one another to form a double helix containing ten bases per helical turn. Each strand of the DNA molecule contains a sequence of nucleotides which is exactly *complementary* to the nucleotide sequence of its partner strand.

The formation of nucleotides in a DNA strand can be decoded for the proteins' synthesis in a living organism. There is a correspondence between the binary format of the nucleotides and the 20 amino acids, which are available to synthesize proteins. This correspondence is described as the *gene expression*. It is used when a cell converts a nucleotide sequence from a gene into the amino acid sequence of a protein.

A *genome* is the complete set of information in the DNA of an organism. Each human cell contains 2 m of DNA strands, which are tucked into a 5–8 μm diameter cell nucleus (Fig. 5.29). Cells, by generating a series of coils and loops, pack these strands into *chromosomes*. The *human genome* consists of approximately 3.2×10^9 nucleotides distributed over 24 chromosomes.

A *Chromatin* is a complex of DNA and proteins. Except in the germ cells (sperm and eggs) and highly specialized eukaryotic cells (such as red blood cells) all other human cells contain two copies of such a complex. One is inherited from the mother and one is inherited from the father and they are called *chromosomes*.

Chromosomes carry genes as codes embedded into their different sites. In general, the more complex the organism the larger is its genome. The human genome is about 200 times larger than that of yeast. However, this is not always true; amoeba's genome is about 200 times larger than that of a human! In addition to genes, a large excess of interspersed DNA is available, which does not seem to carry any useful information. These excessive portions may have appeared in a long-term evolutionary process of different species. However, the real applications of interspersed DNA are still debatable.

DNA replication is the duplication process of DNA, which occurs before a cell divides into two daughter cells which are genetically identical. The replication process needs elaborate consequence of functions by the organelles and proteins within a cell. In addition, specialized enzymes from a cell are needed for carrying out repairs when the components of a cell are damaged during the replication process. Despite all the preemptive measures by a cell, permanent damage may still occur during the replication. Such changes from the norm are referred to as *mutations*. The mutations are often detrimental, having the ability to cause genetic diseases such as many types of cancers. Nevertheless, mutations can be advantageous as well. For instance, bacteria can make their next generation resistant to antibiotics.

Base-pairing is fundamental to the DNA replication process. For the synthesis of a new complementary DNA strand, the initial strand functions as a template. In order to produce these single strands, the double helix must unwind. The hydrogen bonds make the DNA double helix a stable structure, so sufficient energy is needed to separate these bonds and can be applied using exposure to mechanical energy, thermal energy, and radiation by electromagnetic waves.

The process of DNA replication is initiated by *initiator proteins* in a cell. They first bind to the DNA and then separate the two strands apart. Firstly, these proteins separate a short length of the strand at room temperature at a segment which is called the *replication origin*. After starting the process, the initiator protein attracts other proteins, whose duty is to continue the replication process.

The replication is a bidirectional process which has a rate of replication in the order of 100 nucleotide pairs per second. *DNA polymerase*, which is the main protein in synthesizing a DNA, catalyzes the addition of nucleotides to the 3′ end. It forms a *phosphodiester bond* between this end and the 5′-phosphate group of the incoming nucleotide. The energy-rich nucleotide triphosphate provides the energy for the polymerization, which links the nucleotide monomers to the chain and releases pyrophosphate (PP_i). Pyrophosphate is further hydrolyzed to inorganic phosphate (P_i) which makes the polymerization reaction irreversible.

5.5.2   The Structure of RNA

For making a new strand of DNA, *ribonucleic acid* (*RNA*) of a length of ten nucleotides long is used as the *primer*. *Primase* is the enzyme, which uses RNA primer to synthesize a strand of DNA. A strand of RNA is similar to a strand of DNA except that it is made of *ribonucleotide* subunits. In RNA the sugar is ribose and not deoxyribose as in DNA. RNA also has another difference as well: the base thymine (T) is replaced by the base *uracil* (*U*).

Three more enzymes are needed to produce a continuous new DNA strand from pieces of RNA and DNA. They are *nuclease*, which removes the RNA primer, *repair polymerase* that replaces it with DNA, and *DNA ligase* that joins the DNA fragments together. Eventually in the replication process, *telomerase* puts in the ends of the eukaryotic chromosomes.

Cells also have an extra protein to reduce the occurrence of errors in the DNA replication which is called *DNA mismatch repair*. DNA is continually undergoing thermal collisions with other molecules, which results in changes and thus damage to the DNA. In addition, amino groups may be lost from cytosine. UV light is also another source of damage as it may produce a *thymine dimer* by covalently linking two adjacent pyrimidine bases.

5.5.3   DNA Decoders and Microarrays

These days, new methods for analyzing and manipulating DNA, RNA, and proteins are fuelling an information explosion. By knowing the sequence of the nucleotides in a bio-structure, we can obtain the genetic blueprints organisms [6].

In the 1970s it became possible to isolate pieces of DNA in chromosomes. It is now possible to generate new DNA molecules and introduce them back into living organisms. This process is referred to with different expressions such as *recombinant DNA*, *gene splicing*, and *genetic engineering*. Using the process, we can create chromosomes with combinations of genes, which are not present in nature.

In sensing applications, the recombinant DNA techniques can be utilized for revealing the relationships between phenomena in cells, detecting the mutations in DNA, which are responsible for inherited diseases, and identifying possible suspects.

Recombinant DNA techniques are used for unlocking the genome by breaking large pieces of DNA into smaller pieces. Codes in the genome are embedded within these fragments and so by studying these fragments, we can gain an understanding of which proteins the DNA code for.

A class of bacterial enzymes known as *restriction nucleases* is used for cutting DNA at particular sites. These sites are identified by a short sequence of nucleotide pairs. Restriction nucleases are now commonly used in DNA technology, with hundreds of them available in the market. Each one is able to cut DNA at a particular site, giving researchers powerful tools to investigate DNA-related areas.

Gel electrophoresis is one of the first procedures that were utilized to separate cleaved DNA pieces. The most common gels are made of agarose. In the *electrophoresis* processes, after gene splicing, the mixture of DNA fragments is loaded at one side of the gel slab and a voltage is applied (Fig. 5.32). The fragments of DNA travel with velocities, which are proportional to their mass and consequently to their sizes. As a result, the fragments become separated according to their sizes. This produces a pattern that can be decoded and used as the expression of a gene.

The two strands of DNA are held together by hydrogen bonds that can be broken at temperatures higher than 90 °C or at extreme pHs. If the process is reversed the complementary strands will re-form double helices in a process called *hybridization*.

The hybridization of DNA is used for diagnostic applications. For this propose, *DNA probes* are used. A DNA probe is a single-stranded DNA molecule, typically 10–1,000 nucleotides long. Those probes are used in detecting nucleic acid molecules, which contain complementary sequences.

One of the most important applications of DNA probes in diagnostics is for identifying the carriers of genetic diseases. For instance, for *sickle-cell anemia* the exact nucleotide change in a mutant has been determined, which can be used in the sensing process. If the sequence G-A-G is changed to G-T-G at certain positions of the DNA strand then the person is diagnosed with the disease.

A common laboratory procedure utilized for hybridization in conjunction with electrophoresis for the decoding of DNA fragments is called *Southern blotting* (Fig. 5.33). In Southern blotting the unlabelled fragments of DNA, which are separated using electrophoresis, are transformed onto a nitrocellulose paper and then probed with a known gene or fragment for decoding.

DNA microarrays, developed in the last three decades, have revolutionized the way we analyze genes, by allowing the DNA fragments and the RNA products of thousands of genes to be investigated simultaneously.

Using microarrays, the cellular physiology in terms of the *gene expression* can be patterned. The mechanisms that control gene expression act as both an on/off switch

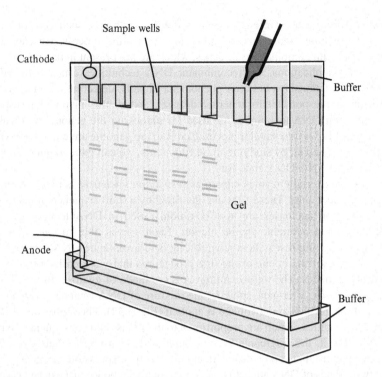

Fig. 5.32 A schematic of gel electrophoresis. After the electrophoresis, in order to visualize the DNA bands, the gel is soaked in a dye which binds to DNA. This particular dye can fluorescence under UV light

to control which genes are expressed in a cell as well as a *volume control* which increases or decreases the level of expression of particular genes if necessary.

DNA microarrays are substrates with a large number of DNA fragments on them in contained areas. Each of these fragments holds a nucleotide sequence that serves as a probe for a specific gene. Some types of microarrays carry DNA fragments corresponding to entire genes that are spotted onto the surface by robots. Others contain short oligonucleotides, which are synthesized on the surface. An *oligonucleotide*, or oligo as it is commonly called, is a short fragment of single-stranded DNA that is typically 5–50 nucleotides long.

In oligonucleotide microarrays, the probes are designed to match known or predicted parts of the sequence of *RNAs*. There are commercially available designs that cover complete genomes from companies such as GE Healthcare, Affymetrix, or Agilent. These microarrays give estimations of the gene expression (Fig. 5.34).

For making the DNA microarray chips, the wafer (which is generally quartz or glass) is patterned with a photoresist and the appropriate mask is put in place. The mask is designed with 1–25 μm^2 windows that allow light to pass through areas, where a specific nucleotide is needed.

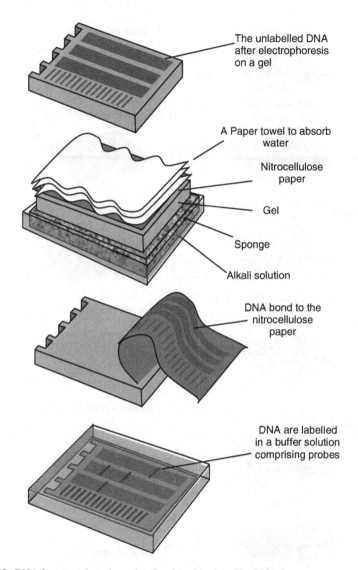

The unlabelled DNA after electrophoresis on a gel

A Paper towel to absorb water

Nitrocellulose paper

Gel

Sponge

Alkali solution

DNA bond to the nitrocellulose paper

DNA are labelled in a buffer solution comprising probes

Fig. 5.33 DNA fragment detection using Southern blotting: The DNA fragments are separated by electrophoresis. A sheet (commonly nitrocellulose and/or nylon) is placed on the gel, and blotting process transfers the DNA fragments. An alkali solution is sucked through the gel and the sheet by a stack of paper towels. The sheet containing the bound single-stranded DNA fragments is removed and placed in a buffer containing the labeled DNA probe. After hybridization, the DNA that has hybridized to the labeled probe shows the code

To use a DNA microarray for monitoring the expression of a gene in a cell, messengerRNA (mRNA), which is the RNA that encodes and carries information from DNA during transcription to the sites of protein synthesis, *is extracted from the cells.* As working with mRNA (mRNA is unstable and is easily degraded by

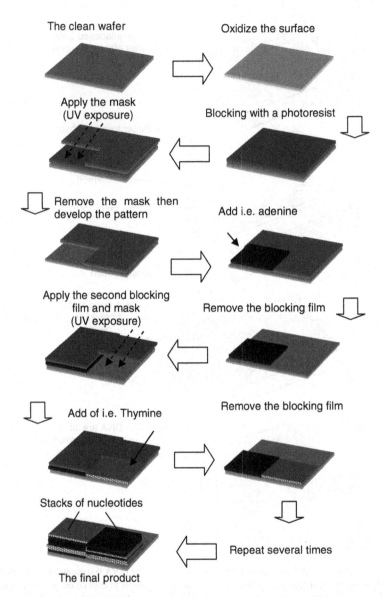

The clean wafer

Oxidize the surface

Apply the mask
(UV exposure)

Blocking with a photoresist

Remove the mask then
develop the pattern

Add i.e. adenine

Apply the second blocking
film and mask
(UV exposure)

Remove the blocking film

Add of i.e. Thymine

Remove the blocking film

Stacks of nucleotides

Repeat several times

The final product

Fig. 5.34 The fabrication of a DNA microarray through the use of photolithography and combi-natorial chemistry-specific probes

RNases which can be found even on our skin) is difficult, an enzyme called *reverse transcriptase* that produces a DNA copy (*complementary DNA or cDNA*) of each mRNA strand is used instead. *The cDNA is also easier to manipulate than the original RNA. The cDNA can be labeled with a fluorescent probe.*

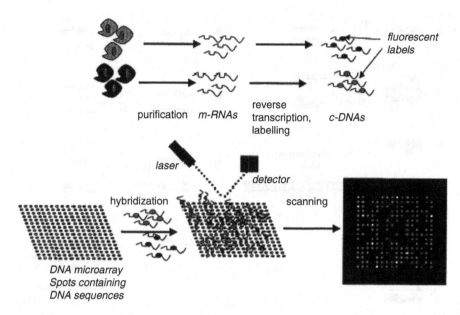

Fig. 5.35 A typical gene microarray experiment. In this example, a comparative gene expression experiment is shown (reprinted from [8] with permission)

The microarray is incubated with the labeled cDNA sample, and hybridization occurs. The array is then washed to remove the unbound molecules, and a scanning laser microscope can be used to find the fluorescent spots. A typical gene microarray-based experiment is shown in Fig. 5.35. The array is optically scanned and the scanned image is then interpreted. The final image is of spots of differing intensity of the label (shown by the brightness of the fluorescent labels) that are related to the degree of hybridization between the probe and the target DNA [8].

The array positions are then compared to particular genes whose expressions are known. The cDNAs which are the reverse of the transcribed mRNAs can be found in sets of libraries. A *cDNA library* refers to a nearly complete set of all the mRNAs contained within a cell or organism. Such a library has several uses: it is important for analyses in *bioinformatics*; in addition, the cDNA sequence gives the genetic relationship between organisms through the similarity of their cDNA. By using these data, novel DNA molecules can be engineered.

The example shown in Fig. 5.35 is a comparative gene expression experiment. Two sets of cells, for example, diseased and healthy cells, provide the initial sample. In the first stage of the experiment, mRNAs are extracted from the cells and reverse transcribed into the more stable cDNA. The cDNAs from each cell population are then labeled, typically with different colored fluorescent dyes. The cDNAs are then hybridized to a DNA microarray and the hybridized array is washed and scanned. The positions and intensities of the spots seen provide information regarding the genes, which are expressed by the cell.

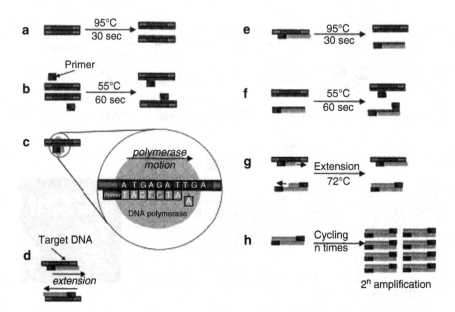

Fig. 5.36 A schematic diagram of the PCR cycle: (a) denaturing, (b) annealing, (c–d) elongation, (e–g) repeat, (h) overall outcome of the process after n times cycling (reprinted from [8] with permission)

Generally, for DNA tests it is necessary to amplify the number of DNA strands to a sufficient quantity. The *polymerase chain reaction (PCR)* is a powerful form of DNA amplification for this purpose. The PCR process usually consists of a series (up to 35 cycles), each of which consists of three steps (Fig. 5.36):

(a) The double-stranded DNA is heated to 94–96 °C in order to separate the strands. This step is called *denaturing*; it breaks apart the hydrogen bonds that connect the two DNA strands.

(b) After separating the DNA strands, the temperature is lowered so that the *artificial primers* (often not more than 50 and usually only 18–25 base pairs long nucleotides that are complementary to the beginning and the end of the DNA fragment to be amplified) can attach themselves to the single DNA strands. This step is called *annealing*. The temperature of this stage depends on the primers being used (45–60 °C).

(c and d) Finally, the DNA polymerase protein has to copy the DNA strands. It starts at the annealed primer and works its way along the DNA strand in a step called *elongation*. This produces a double strand from the single-strand DNA.

The rapid evolution of nucleic acid based assays in the form of DNA chips is one of the latest developments in biosensor industry. The concept of a million hybridization assays performed simultaneously on a one-square centimeter chip has much in common with the ultimate goal of high-density sensor arrays.

   In addition to DNA microarrays, a large number of other DNA-based biosensors have also been described in the literature with electrochemical, optical, and piezoelectric transducing elements. Such biosensors generally take advantage of the DNA strands' lock and key properties to achieve high selectivity.

   Basically most of the recent biochips, which were described in the previous section, are optical sensors as the DNA can be readily labeled with optical tags. These tags are detected using the irradiation and measurements of the absorbed light.

   DNA structure can be used to develop electronic sensing devices. Charge carriers can hop along the DNA over distances of at least a few nanometers and in this way DNA can act as a *molecular wire*. As a result, DNA systems represent a novel base material for sensor templates.

5.6   Summary

The final chapter of this book focused on organic sensors: sensors which either incorporate organic materials in their structures or are used for sensing organic materials.

   To understand the behavior and functionality of such sensors, the interaction of different types of surfaces with organic materials were described and methods for the fictionalizations of sensors' surfaces were presented. The emphasis was on materials such as gold, carbon, silicon, metal oxides, and polymers, which are the most common materials in the fabrication of biosensors.

   The definitions of the most of commonly employed biomaterials including proteins and DNA were presented and their importance in biosensing applications was illustrated via relevant examples.

References

1. Gizeli E, Lowe CR (2002) Biomolecular sensors. Taylor & Francis, London
2. Zhang J, Wang ZL, Liu J, Chen S, Liu GY (2003) Self-assembled nanostructures. Kluwer, New York
3. Nuzzo RG, Allara DL (1983) Adsorption of bifunctional disulfides on gold surfaces. J Am Chem Soc 105:4481–4483
4. Katsnelson MI (2007) Graphene: carbon in two dimensions. Mater Today 10:20–27
5. Scouten WH (1981) Affinity chromatography: bioselective adsorption on inert matrices. Wiley, New York
6. Alberts B, Bary D, Hopkins K, Johnson A, Lewis J, Raff M, Roberts K, Walter P (2004) Essential cell biology. Garland Science, Oxford
7. Homola J (2003) Present and future of surface Plasmon resonance biosensors. Anal Bioanal Chem 377:528–539
8. Sanders GHW, Manz A (2000) Chip-based microsystems for genomic and proteomic analysis. Trends Anal Chem 19:364–378

Index